This Idea Must Die

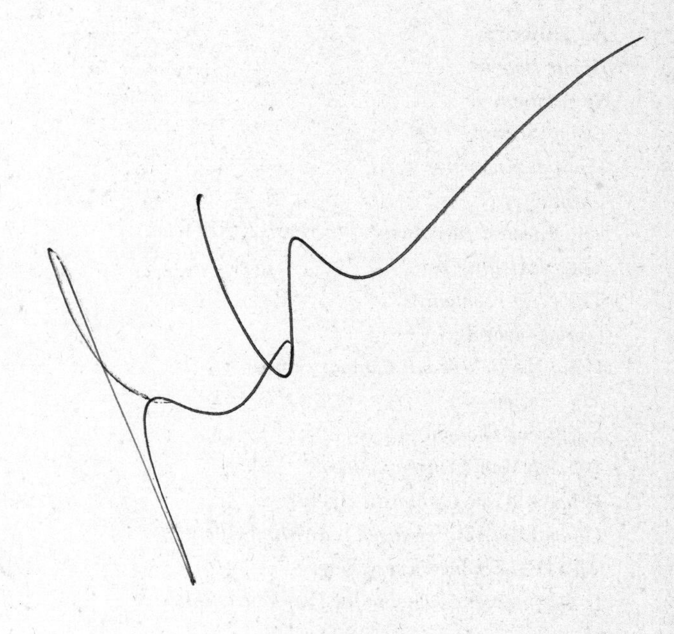

Also by John Brockman

As Author:

By the Late John Brockman
37
Afterwords
The Third Culture: Beyond the Scientific Revolution
Digerati

As Editor:

About Bateson
Speculations
Doing Science
Ways of Knowing
Creativity
The Greatest Inventions of the Past 2,000 Years
The Next Fifty Years
The New Humanists
Curious Minds
What We Believe but Cannot Prove
My Einstein
Intelligent Thought
What Is Your Dangerous Idea?
What Are You Optimistic About?
What Have You Changed Your Mind About?
This Will Change Everything
Is the Internet Changing the Way You Think?
Culture
The Mind
This Will Make You Smarter
This Explains Everything
Thinking
What Should We Be Worried About?
The Universe

As Coeditor:

How Things Are (with Katinka Matson)

This Idea Must Die

Scientific Theories That Are Blocking Progress

Edited by John Brockman

HARPER PERENNIAL

NEW YORK • LONDON • TORONTO • SYDNEY • NEW DELHI • AUCKLAND

HARPER PERENNIAL

FIRST EDITION

Designed by William Ruoto

Library of Congress Cataloging-in-Publication Data is available upon request.

ISBN 978-0-06-237434-9

17 18 19 OV/RRD 10 9

To Richard Dawkins, Daniel C. Dennett,
Jared Diamond, and Steven Pinker

Pioneers of the Third Culture

CONTENTS

ACKNOWLEDGMENTS

My thanks to Laurie Santos for suggesting this year's *Edge* Question and to Paul Bloom and Jonathan Haidt for their refinements. As always, thanks to Stewart Brand, Kevin Kelly, George Dyson, and Steven Pinker, for their continued support. I also wish to thank Peter Hubbard of HarperCollins for his encouragement. I am also indebted to my agent, Max Brockman, who saw the potential for this book, and, as always, to Sara Lippincott for her thoughtful and meticulous editing.

JOHN BROCKMAN
Publisher & Editor, Edge

THE 2014 *EDGE* QUESTION

Science advances by discovering new things and developing new ideas. Few truly new ideas are developed without abandoning old ones first. As theoretical physicist Max Planck (1858–1947) noted, "A new scientific truth does not triumph by convincing its opponents and making them see the light, but rather because its opponents eventually die, and a new generation grows up that is familiar with it." In other words, science advances by a series of funerals. Why wait that long?

WHAT SCIENTIFIC IDEA IS READY FOR RETIREMENT?

Ideas change, and the times we live in change. Perhaps the biggest change today is the rate of change. What established scientific idea is ready to be moved aside so that science can advance?

THE THEORY OF EVERYTHING

GEOFFREY WEST

Theoretical physicist; Distinguished Professor and past president, Santa Fe Institute

Everything? Well, wait a minute. Questioning a Theory of Everything may be beating a dead horse, since I'm certainly not the first to be bothered by its implicit hyperbole, but let's face it, referring to one's field of study as "the Theory of Everything" smacks of arrogance and naïveté. Although it's been around for only a relatively short period and may already be dying a natural death, the phrase (though certainly not the endeavor) should be retired from serious scientific literature and discourse.

Let me elaborate. The search for grand syntheses, for commonalities, regularities, ideas, and concepts transcending the narrow confines of specific problems or disciplines is one of the great inspirational drivers of science and scientists. Arguably, it is also a defining characteristic of *Homo sapiens sapiens*. Perhaps the binomial form of *sapiens* is some distorted poetic recognition of this. Like the invention of gods and God, the concept of a Theory of Everything connotes the grandest vision of all, the inspiration of all inspirations: namely, that we can encapsulate and understand the entirety of the universe in a small set of precepts—in this case, a concise set of mathematical equations. Like the concept of God, however, it's potentially misleading and intellectually dangerous.

Among the classic grand syntheses in science are Newton's laws, which taught us that heavenly laws were no different than the earthly; Maxwell's unification of electricity and magnetism, which brought the ephemeral aether into our lives; Darwin's theory of natural selection, which reminded us that we're just animals and plants after all; and the laws of thermodynamics, which suggest we can't go on forever. Each of these has had profound consequences—not only in changing the way we think about the world but also in laying the foundations for technological advancements that have led to the standard of living many of us are privileged to enjoy. Nevertheless, they're all, to varying degrees, incomplete. Indeed, understanding the boundaries of their applicability and the limits to their predictive power, and the ongoing search for exceptions, violations, and failures, have provoked even deeper questions and challenges, stimulating the continued progress of science and the unfolding of new ideas, techniques, and concepts.

One of the great scientific challenges is the search for a Grand Unified Theory of the elementary particles and their interactions, including its extension to understanding the cosmos and even the origin of spacetime itself. Such a theory would be based on a parsimonious set of underlying mathematizable universal principles that integrate and explain all the fundamental forces of nature, from gravity and electromagnetism to the weak and strong nuclear forces, incorporating Newton's laws, quantum mechanics, and general relativity. Fundamental quantities like the speed of light, the dimensionality of spacetime, and the masses of the elementary particles would all be predicted, and the equations governing the origin and evolution of the universe through to the formation of galaxies and beyond would be derived—and so on. This constitutes the Theory of Everything. It's a truly remarkable and enormously ambitious quest, which has occupied thousands of researchers for over fifty years at a cost of billions of dollars. Measured by almost any metric, this quest, which is still far from its ultimate goal, has been enormously successful, leading, for example, to the discovery of quarks and the Higgs boson, to black holes and the Big Bang, to quantum chromodynamics and string theory . . . and to many Nobel Prizes.

But "Everything"? Well, hardly. Where's life, where are animals and cells, brains and consciousness, cities and corporations, love and hate, etc., etc.? How does the extraordinary diversity and complexity seen here on Earth arise? The simplistic answer is that these are inevitable outcomes of the interactions and dynamics encapsulated in the Theory. Time evolves from the geometry and dynamics of strings, the universe expands and cools, and the hierarchy—from quarks to nucleons, to atoms and molecules, to cells, brains, and emotions and all the rest—comes tumbling out, a sort of *deus ex machina*, a result of "just" turning the crank of increasingly complicated equations and computations presumed, in principle, to be soluble to any sufficient degree of accuracy. Qualitatively, this extreme version of reductionism may have some validity, but Something is missing.

The "Something" includes concepts like information, emergence, accidents, historical contingency, adaptation, and selection—all characteristics of complex adaptive systems, whether organisms, societies, ecosystems, or economies. These are composed of myriad individual constituents or agents, which take on collective characteristics generally unpredictable (certainly in detail) from their underlying components, even if the interactive dynamics are known. Unlike the Newtonian paradigm on which the Theory of Everything is based, the complete dynamics and structure of complex adaptive systems cannot be encoded in a small number of equations. Indeed, in most cases, probably not even in an infinite number! Furthermore, predictions to arbitrary degrees of accuracy are not possible, even in principle.

Perhaps, then, the most surprising consequence of a visionary Theory of Everything is that it implies that on the grand scale the universe—including its origins and evolution—although extremely complicated, is not complex but in fact surprisingly simple, since it can be encoded in a limited number of equations. Conceivably, only one. This is in stark contrast to what we encounter here on Earth, where we're integral to some of the most diverse, complex, and messy phenomena occurring anywhere in the universe, and which require additional, possibly nonmathematizable, concepts to understand. So, while applauding and admiring the search for a Grand Unified Theory of all the basic forces of nature, let's drop the implication that it can in principle explain and predict *Everything.* Let us instead incorporate a parallel quest for a Grand Unified Theory of Complexity. The challenge of developing a quantitative, analytic, principled, predictive framework for understanding complex adaptive systems is surely a grand challenge for the 21st century. Like all grand syntheses, it will inevitably remain incomplete, but nevertheless it will undoubtedly inspire significant, possibly revolutionary new ideas, concepts, and techniques.

UNIFICATION

MARCELO GLEISER
Theoretical physicist, Dartmouth College; author,
The Island of Knowledge; The Limits of Science and the Search for Meaning

There! I said it! The venerable notion of Unification needs to go. I don't mean the smaller unifications we scientists search for all the time, connecting as few principles with as many natural phenomena as possible. This sort of scientific economy is a major foundational stone for what we do: We search and we simplify. Over the centuries, scientists have done wonders following this motto. Newton's law of universal gravity, the laws of thermodynamics, electromagnetism, universal behavior in phase transitions. . . .

The trouble starts when we take this idea too far and search for the *Über*-unification, the Theory of Everything, the arch-reductionist notion that all forces of nature are merely manifestations of a single force. This is the idea that needs to go. And I say this with a heavy heart; my early career aspirations and formative years were very much fueled by the impulse to unify it all.

The idea of unification is quite old, as old as Western philosophy. Thales, the first pre–Socratic philosopher, posited that "All is water," thus dreaming up a single material principle to describe all of nature. Plato proposed elusive geometrical forms as the archetypal structures behind all there is. Math became equated with beauty and beauty with truth. From there, the highest of post–Plato aspirations was to erect a purely mathematical explanation for all there is: the all-encompassing cosmic blueprint, the masterwork of a supreme intelligence. Needless to say, the whole thing was always about our intelligence, even if often attributed to some foggy "mind of God" metaphor. We explain the world the way we think about it. There's no way out of our minds.

The impulse to unify it all runs deep in the souls of mathematicians and theoretical physicists, from the Langlands program to superstring theory. But here's the rub: Pure mathematics isn't physics. The power of mathematics comes precisely from its detachment from physical reality. A mathematician can create any universe she wants and play all sorts of games with it. A physicist can't; his job is to describe nature as we perceive it. Nevertheless, the unification game has been an integral part of physics since Galileo and has produced what it should: approximate unifications.

Yes, even the most sacred of our unifications are only approximations. Take, for example, electromagnetism. The equations describing electricity and magnetism are perfectly symmetric only in the absence of any sources of charge or magnetism—that is, in empty space. Or take the famous (and beautiful) Standard Model of particle physics, based on the "unification" of electromagnetism and the weak nuclear force. Here again, we don't have a real unification, since the theory retains two forces all along. (In more technical jargon, there are two coupling constants and two gauge groups.) A real unification, such as the conjectured Grand Unification between the strong, the weak, and the electromagnetic forces, proposed forty years ago, remains unfulfilled.

So, what's going on? Why do so many insist in finding the One in Nature while Nature keeps telling us that it's really about the Many?

For one thing, the scientific impulse to unify is crypto-religious. The West has bathed in monotheism for thousands of years, and even in polytheistic cultures there's always an alpha God in charge (Zeus, Ra, Para-Brahman). For another, there's something deeply appealing in equating all of nature to a single creative principle: To decipher the "mind of God" is to be special, is to answer to a higher calling. Pure mathematicians who believe in the reality of mathematical truths are monks of a secret order, open only to the initiated. In the case of high energy physics, all unification theories rely on sophisticated mathematics related to pure geometric structures: The belief is that nature's ultimate code exists in the ethereal world of mathematical truths and that we can decipher it.

Recent experimental data has been devastating to such belief—no trace of supersymmetric particles, of extra dimensions, or of dark matter of any sort, all long-awaited signatures of unification physics. Maybe something will come up; to find, we must search. The trouble with unification in high energy physics is that you can always push it beyond the experimental range. "The Large Hadron Collider got to 7 TeV and found nothing? No problem! Who said nature should opt for the simplest versions of unification? Maybe it's all happening at much higher energies, well beyond our reach."

There's nothing wrong with this kind of position. You can believe it until you die, and die happy. Or you can conclude that what we do best is construct approximate models of how nature works and that the symmetries we find are only descriptions of what really goes on. Perfection is too hard a burden to impose on nature.

People often see this kind of argument as defeatist, as coming from someone who got frustrated and gave up. (As in "He lost his faith.") Big mistake. To search for simplicity is essential to what scientists do. It's what I do. There are essential organizing principles in nature, and the laws we find are excellent ways to describe them. But the laws are many, not one. We're successful pattern-seeking rational mammals. That alone is cause for celebration. However, let's not confuse our descriptions and models with reality. We may hold perfection in our mind's eye as a sort of ethereal muse. Meanwhile nature is out there doing its thing. That we manage to catch a glimpse of its inner workings is nothing short of wonderful. And that should be good enough.

SIMPLICITY

A. C. GRAYLING

Philosopher; founder and master, New College of the Humanities, London; supernumerary fellow, St. Anne's College, Oxford; author, The Good Argument: The Case Against Religion and for Humanism

When two hypotheses are equally adequate to the data, and equal in predictive power, extratheoretical criteria for choosing between them might come into play. They include not just questions about best fit with other hypotheses or theories already predicated to inquiry but also the aesthetic qualities of the competing hypotheses themselves—which is more pleasing, more elegant, more beautiful?—and of course the question of which of them is simpler.

Simplicity is a desideratum in science, and the quest for it is a driver in the task of effecting reductions of complex phenomena to their components. It lies behind the assumption that there must be a single force in nature, of which the gravitational, electroweak, and strong nuclear forces are merely manifestations; and this assumption in turn is an instance of the general view that there might ultimately be a single kind of thing (or stuff or field or as-yet-undreamt-of phenomenon) out of which variety springs by means of principles themselves fundamental and simple.

Compelling as the idea of simplicity is, there's no guarantee that nature itself has as much interest in simplicity as those attempting to describe it. If the idea of emergent properties still has purchase, biological entities cannot be fully explained except in terms of them, which means in their full complexity, even though considerations of structure and composition are indispensable.

Two measures of complexity are the length of the message required to describe a given phenomenon and the length of the evolutionary history of that phenomenon. On a certain view, that makes a Jackson Pollock painting complex by the first measure, simple by the second, whereas a smooth pebble on a beach is simple by the first and complex by the second. The simplicity sought in science might be thought to be what is achieved by reducing the length of the descriptive message—encapsulation in an equation, for example. But could there be an inverse relationship between the degree of simplicity achieved and the degree of approximation that results?

Of course it would be nice if everything in the end turned out to be simple, or could be made amenable to simple description. But some things might be better or more adequately explained in their complexity—biological systems again come to mind. Resisting too dissipative a form of reductionism might ward off those silly kinds of criticism claiming that science aims to see nothing in the pearl but the disease of the oyster.

THE UNIVERSE

SETH LLOYD

Professor of quantum mechanical engineering, MIT; author, Programming the Universe

I know. The universe has been around for 13.8 billion years and is likely to survive for another 100 billion years or more. Plus, where would the universe retire to? Florida isn't big enough. But it's time to retire the 2,500-year-old scientific idea of the universe as the single volume of space and time that contains everything. Twenty-first-century cosmology strongly suggests that what we see in the cosmos—stars, galaxies, space and time since the Big Bang—does not encompass all of reality. Cosmos, buy the condo.

What is the universe, anyway? To test your knowledge of the universe, please complete the following sentence. The universe

(a) consists of all things visible and invisible—what is, has been, and will be.
(b) began 13.8 billion years ago in a giant explosion called the Big Bang and encompasses all planets, stars, galaxies, space, and time.
(c) was licked out of the salty rim of the primordial fiery pit by the tongue of a giant cow.
(d) All of the above.
(Correct answer below.)

The idea of the universe as an observed and measured thing has persisted for thousands of years. Those observations and measurements have been so successful that today we know more about the origin of the universe than we do about the origin of life on Earth. But the success of observational cosmology has brought us to a point where it's no longer possible to identify the universe—in the sense of answer (a) above—with the observed cosmos—answer (b). The same observations that establish the detailed history of the universe imply that the observed cosmos is a vanishingly small fraction of an infinite universe. The finite amount of time since the Big Bang means that our observations extend only a little more than 10 billion light-years from Earth. Beyond the horizon of our observation lies more of the same—space filled with galaxies stretching on forever. No matter how long the universe exists, we will have access to only a finite part, while an infinite amount of universe remains beyond our ken. All but an infinitesimal fraction of the universe is unknowable.

That's a blow. The scientific concept *universe = observable universe* has thrown in the towel. Perhaps that's OK. What's not to like about a universe encompassing infinite unknowable space? But the hits keep coming. As cosmologists delve deeper into the past, they find more and more clues that, for better or worse, there's more out there than just the infinite space beyond our horizon. Extrapolating backward in time to the Big Bang, cosmologists have identified an epoch called inflation, in which the universe doubled in size many times over a tiny fraction of a second. The vast majority of spacetime consists of this rapidly expanding stuff. Our own universe, infinite as it is, is just a "bubble" that has nucleated in this inflationary sea.

It gets worse. The inflationary sea contains an infinity of other bubbles, each an infinite universe in its own right. In different bubbles, the laws of physics can take different forms. Somewhere out there in another bubble universe, the electron has a different mass. In another bubble, electrons don't exist. Because it consists not of one cosmos but of many, the multi-bubble universe is often called a multiverse. The promiscuous nature of the multiverse may be unappealing (William James, who coined the word, called the multiverse a "harlot"), but it's hard to eliminate. As a final insult to unity, the laws of quantum mechanics indicate that the universe is continually splitting into multiple histories, or "many worlds," out of which the world we experience is only one. The other worlds contain the events that didn't happen in our world.

After a two-millenium run, the universe as observable cosmos is kaput. Beyond what we can see, an infinite array of galaxies exists. Beyond that infinite array, an infinite number of bubble universes bounce and pop in the inflationary sea. Closer by, but utterly inaccessible, the many worlds of quantum mechanics branch and propagate. MIT cosmologist Max Tegmark calls these three kinds of proliferating realities the type I, type II, and type III multiverses. Where will it all end? Somehow, a single, accessible universe seemed more dignified.

There's hope, however. Multiplicity itself represents a kind of unity. We now know that the universe contains more things than we can ever see, hear, or touch. Rather than regarding the multiplicity of physical realities as a problem, let's take it as an opportunity.

Suppose that everything that could exist does exist. The multiverse is not a bug but a feature. We have to be careful: The set of everything that could exist belongs to the realm of metaphysics rather than physics. Tegmark and I have shown that with a minor restriction, however, we can pull back from the metaphysical edge. Suppose that the physical multiverse contains all things that are locally finite, in the sense that any finite piece of the thing can be described by a finite amount of information. The set of locally finite things is mathematically well defined: It consists of things whose behavior can be simulated on a computer (more specifically, on a quantum computer). Because they're locally finite, the universe we observe and the various other universes are all contained within this computational universe. As is, somewhere, a giant cow.

Answer to quiz: (c)

IQ

SCOTT ATRAN

Anthropologist, Centre Nationale de la Recherche Scientifique, Paris; author, Talking to the Enemy: Religion, Brotherhood, and the (Un)Making of Terrorists

There's no reason to believe, and much reason not to believe, that the measure of a so-called "Intelligence Quotient" in any way reflects some basic cognitive capacity or "natural kind" of the human mind. The domain-general measure of IQ isn't motivated by any recent discovery of cognitive or developmental psychology. It thoroughly confounds domain-specific abilities—distinct mental capacities for, say, geometrical and spatial reasoning about shapes and positions, mechanical reasoning about mass and motion, taxonomic reasoning about biological kinds, social reasoning about other people's beliefs and desires, and so on—which are the only sorts of cognitive abilities for which an evolutionary account seems plausible in terms of natural selection for task-specific competencies.

Nowhere in the animal or plant kingdoms does there ever appear to have been natural selection for a task-general adaptation. An overall measure of intelligence or mental competence is akin to an overall measure for "the body," taking no special account of the various and specific bodily organs and functions, such as hearts, lungs, stomach, circulation, respiration, digestion, and so on. A doctor or biologist presented with a single measure for "Body Quotient" (BQ) wouldn't be able to make much of it.

IQ is a general measure of socially acceptable categorization and reasoning skills. IQ tests were designed in behaviorism's heyday, when there was little interest in cognitive structure. The scoring system was tooled to generate a normal distribution of scores with a mean of 100 and a standard deviation of 15.

In other societies, a normal distribution of some general measure of social intelligence might look very different; some "normal" members of our society could well produce a score that's a standard deviation from "normal" members of another society on that other society's test. For example, in forced-choice tasks, East Asian students (Chinese, Koreans, Japanese) tend to favor field-dependent perception over object-salient perception, thematic reasoning over taxonomic reasoning, and exemplar-based categorization over rule-based categorization. American students generally prefer the opposite. On tests measuring these various categorization and reasoning skills, East Asians average higher on their preferences and Americans average higher on theirs. There's nothing particularly revealing about these different distributions other than that they reflect underlying sociocultural differences.

There's a long history of acrimonious debate over which, if any, aspects of IQ are heritable. The most compelling studies concern twins raised apart and adoptions. Twin studies rarely have large sample populations; moreover, they often involve twins separated at birth because a parent dies or cannot afford to support both and one is given over to be raised by relatives, friends, or neighbors. This disallows ruling out the effects of social environment and upbringing in producing convergence among the twins. The chief problem with adoption studies is that the mere fact of adoption reliably increases IQ, regardless of any correlation between the IQs of the children and those of their biological parents. Nobody has the slightest causal account of how or why genes, singly or in combination, might affect IQ. I don't think it's because the problem is too hard but because IQ is a specious rather than a natural kind.

BRAIN PLASTICITY

LEO M. CHALUPA
Vice president of research, George Washington University

Brain plasticity refers to the ability of neurons to change their structural and functional properties with experience. That hardly seems surprising, since every part of the body changes with age. What's special about brain plasticity (but not unique to this organ) is that the changes are mediated by events that are in some sense adaptive. The field of brain plasticity primarily derives from the pioneering studies of Torsten Wiesel and David Hubel, who showed that depriving one eye of normal visual input during early development resulted in a loss of functional connections of that eye with the visual cortex, while the connections of the eye not deprived of visual input expanded.

These studies convincingly demonstrated that early brain connections aren't hardwired but can be modified by early experience; hence they're plastic. For this work and related studies done in the 1960s, Wiesel and Hubel received the 1981 Nobel Prize in physiology or medicine. Since that time, there have been thousands of studies showing a wide diversity of neuronal changes in virtually every region of the brain, ranging from the molecular to the systems level, in young, adult, and elderly subjects. As a result, by the end of the 20th century our view of the brain had evolved from the hardwired to the seemingly ever changeable. Today, "plasticity" is one of the most commonly used words in the neuroscience literature. Indeed, I've employed this term many times in my own research articles and used it in the titles of some of my edited books. So what's wrong with that, you may ask?

For one thing, the widespread application of "brain plasticity" to virtually every type of change in neuronal structure and function has rendered the term largely meaningless. When almost any change in neurons is characterized as plasticity, the term encompasses so much that it no longer conveys any useful information. Moreover, many studies invoke brain plasticity as the underlying cause of modified behavioral states without having any direct evidence for neuronal changes. Particularly egregious are the studies showing improvement in performance on some particular task with practice. The fact that practice improves performance was noted before anything was known about the brain. Does it really add anything to state that improvements in function demonstrate a remarkable degree of brain plasticity. The word "remarkable" is often used to denote practice effects in seniors, as if those old enough to receive Social Security are incapable of showing enhanced performance with training.

Studies of this type have led to the launch of the brain-training industry. Many of these programs are focused on the very young. Particularly popular in past years was the "Mozart effect," which led parents without any interest in classical music themselves to continually play pieces by Mozart to their infants. This movement seems to have abated, replaced by a plethora of games supposed to improve the brains of children of all ages. But the largest growth in the brain-plasticity industry has focused on the aging brain. Given the concerns most of us have about memory loss and decreasing cognitive abilities with age, this is understandable. There are large profits to be made, as is evident from the number of companies that have proliferated in this sector in recent years.

There is of course nothing wrong in having children or seniors engage in activities that challenge their cognitive functions. In fact, there may be some genuine benefits in doing so. Certainly, undergoing such training is preferable to watching television for hours each day. It's also the case that any and all changes in performance reflect some underlying changes in the brain. How could it be otherwise, since the brain controls all behaviors? But as yet we don't know what occurs in the brain when performance on a specific video game improves, nor do we understand how to make such changes long-lasting and generalizable to diverse cognitive states. Terming such efforts "brain training" or "enhanced brain plasticity" is often just hype intended to sell a product. This doesn't mean the so-called brain exercises should be abandoned; they're unlikely to cause harm and may even do some good. But please refrain from invoking brain plasticity, remarkable or otherwise, to explain the resulting improvements.

CHANGING THE BRAIN

HOWARD GARDNER

Hobbs Professor of Cognition and Education, Harvard Graduate School of Education; author, Truth, Beauty, and Goodness Reframed

When I speak to students or lay audiences about any kind of digital innovation, the feedback from my listeners goes, roughly, "Do smartphones change the brain?" or "We can't let infants play with pads, because it might affect their brains." I try to explain that *everything* we do affects the nervous system, and that their comments are therefore either meaningless or need to be unpacked.

Here is one such unpacking: "Does this experience affect the nervous system significantly and perhaps even permanently?" Another: "Do you mean 'affect the mind' or 'affect the brain'?"

When the questioner looks blank, I sense that he or she needs a refresher course in philosophy, psychology, and neuroscience.

"THE ROCKET SCIENTIST"

VICTORIA WYATT

Associate professor of indigenous arts of North America, University of Victoria

It's time to retire "the rocket scientist" of cliché fame: *"It doesn't take a rocket scientist to. . . ."*

"The rocket scientist" is a personage rather than a principle, and a fictitious personage at that. He (or she) was constructed by popular usage, not by scientists. Still, the cliché perpetuates outdated public perceptions of scientific principles, and that's critical. "The rocket scientist" needs a good retirement party.

I'll start with a disclaimer. My dreams of that retirement gala may appear tinged with professional envy. I've never heard anyone say, "It doesn't take an ethnohistorian to. . . ." I never will. So, yes, the cliché does slight the humanities, but that's not my concern. Rather, "the rocket scientist," as popularly conceived, dangerously slights the sciences. Our Earth cannot afford that.

"The rocket scientist" stands outside society, frozen on a higher plane. Widely embraced and often repeated, the cliché reflects the public's comfort with divorcing science from personal experience. The cliché imposes a boundary (of brilliance) between the scientist and everyone else. This makes for popular movies and television shows, but it's insidious. Artificially constructed boundaries isolate. They focus attention on differences and distinctions. In contrast, the exploration of relationships and process feeds rapid scientific developments—systems biology, epigenetics, neurology and brain research, astronomy, medicine, quantum physics. Complex relationships also shape the urgent challenges we face: global epidemics, climate change, species extinction, finite resources—these all comprise integral interconnections.

Approaching such problems demands an appreciation of diversity, complexity, relationships, and process. Popular understanding of contemporary science demands the same. We can address urgent global issues only when policy makers see science clearly—when they view diversity, complexity, relationships, and process as essential to understanding rather than as obstacles to it.

At present, though, constructed boundaries pervade not just our clichés but also our institutions and policy structures. Examples abound. Universities segment researchers and students into disciplinary compartments with discrete budgetary line items, competing for scarce resources. ("Interdisciplinary" makes a good buzzword, but the paradigm on which our institutions rest militates against it.) The model of nations negotiating as autonomous entities has failed abysmally to address climate change. In my provincial government's bureaucracy, separate divisions oversee oceans and forests, as if a fatal barrier sliced the ecosystem at the tideline.

Time suffers, too. Past gets alienated from present, and present from future, as our society zooms in on short-term fiscal and political deadlines. Fragmented time informs all other challenges and makes them all the more dire.

So much of our society still operates on a paradigm of simplification, compartmentalization, and boundaries, when we need a paradigm of diversity, complexity, relationships, and process. Our societal structures fundamentally conflict with the messages of contemporary science. How can policy makers address crucial global issues while ignoring contemporary scientific principles?

The real world plays out as a video. The relationships between frames make the story comprehensible. In contrast, "the rocket scientist" stands fictitiously yet firmly alone, on a lofty pinnacle, apart from society, not a part of it. Yes, it's just a cliché, but language matters and jokes instruct. It's time for "the rocket scientist" to retire.

I'll close with another disclaimer: I mean no offense to real rocket scientists. (Some of my best friends have been rocket scientists.) Real rocket scientists exist. They inhabit the real world, with all the attending interconnections, relationships, and complexities. "The rocket scientist" embodies the opposite. We'd all be well served by that retirement.

INDIVI-DUALITY

NIGEL GOLDENFELD

Center for Advanced Study Professor in Physics, director, Institute for Universal Biology, University of Illinois, Urbana-Champaign

In physics, we use the convention that the suffix *-on* denotes a quantized unit of something. For example, in classical physics there are electromagnetic waves. But in the quantum version of the theory, originating with Einstein's 1905 Nobel Prize–winning work, we know that under certain circumstances it's more precise to regard electromagnetic radiant energy as being distributed in particles called *photons*. This wave-particle duality is the underpinning of modern physics—not just photons but a zoo of what were once called elementary particles that include prot*ons*, neutr*ons*, pi*ons*, mes*ons*, and of course the Higgs bos*on*. (Neutrino? It's a long story. . . .)

And what about you? You're a pers*on*. Are you a quantum of something too? Well, clearly there are no fractional humans and we are trivially quantized. But elementary particles, or units, are useful conceptually because they can be considered in isolation, devoid of interactions—like point particles in an ideal gas. You would certainly not fulfill that description, networked, online, and cultured as you undoubtedly are. Your strong interaction with other humans means that your individuality is complicated by the fact that you're part of a society and can function properly only in such a milieu. We could go further and say you're a quantum of a spatially distributed field, one that describes the density around each point in space of humans, rather than the electromagnetic field intensity. This description turns out to be technically powerful for describing the behavior of ecosystems in space and time, particularly for describing extinction, where discontinuous change is important. It seems apt to invoke here the strangely oxymoronic term *indivi-duality*, a counterpart to wave-particle duality.

The notion of "individual" has several other connotations. It can mean discrete or single, but its etymology is also reminiscent of "indivisible." Clearly we're not indivisible but constituted from cells, which are themselves constituted of cytoplasm, nucleic acids, proteins, etc., themselves constituted of atoms, which contain neutrons, protons, electrons, all the way down to the elementary particles now believed to be products of string theory, itself now known not to be a final description of matter. In other words, it's "turtles all the way down," and there are no indivisible units of matter, no meaning to the notion of "elementary particle," no place to stop. Everything is made of something, and so on, ad infinitum.

However, this doesn't mean that everything is simply the sum of its parts. Take the proton, for example—which is made up of three quarks. It has a type of intrinsic angular momentum called spin, which was initially expected to be the sum of that of its constituent quarks. Yet experiments carried out over the last twenty or thirty years have shown that this is not the case: The spin arises out of some shared collective aspect of the quarks and short-lived fluctuating particles called gluons. The notion of individual quarks is not useful when the collective behavior is so strong. The proton is made of something, but its properties aren't found by adding up the properties of its parts. When we try to identify the something, we discover that, as some say of Los Angeles, there's no "there" there.

You probably already knew that naïve reductionism is often too simplistic. However, there's another point. It's not just that you're a composite, something you already knew, but you're in some senses not even human. You have perhaps 100 trillion bacterial cells in your body, numbering 10 times more than your human cells and containing 100 times as many genes as your human cells. These bacteria aren't just passive occupants of the zoo that is you. They self-organize into communities within your mouth, guts, and elsewhere, and these communities—microbiomes—are maintained by varied, dynamic patterns of competition and cooperation between the various bacteria, which allows us to live.

In the last few years, genomics has given us a tool to explore the microbiome by identifying microbes by their DNA sequences. The story emerging from these studies is not yet complete, but it has already led to fascinating insights. Thanks to its microbes, a baby can better digest its mother's milk. And your ability to digest carbohydrates relies to a significant extent on enzymes that can be made only by genes present not in you but in your microbiome. Your microbiome can be disrupted—for example, due to treatment by antibiotics—and in extreme cases can be invaded by dangerous monocultures such as *Clostridium difficile*, leading to your death. Perhaps the most remarkable finding is the gut/brain axis: Your gastrointestinal microbiome can generate small molecules that may be able to pass through the blood-brain barrier and affect the state of your brain. Although the precise mechanism isn't yet clear, there's growing evidence that your microbiome may be a significant factor in mental states such as depression and autism spectrum conditions. In short, you may be a collective property arising from the close interactions of your constitutents.

Now, maybe it's true that you're not an individual in one sense of the word, but how about your microbes? Well, it turns out that your microbes are a strongly interacting system, too: They form dense colonies within you and not only exchange chemicals for metabolism but communicate by emitting molecules. They can even transfer genes between themselves, and in some cases do that in response to signals emitted by a hopeful recipient—a bacterial cry for help! A single microbe in isolation doesn't do these things, thus these complex behaviors are a property of the collective and not the individual microbes. Even microbes that seem to be from the same nominal species can have genomes differing in content by as much as 60 percent of their genes. So much for the intuitive notion of "species"! That's another too-anthropomorphic scientific idea that doesn't apply to most of life.

Up to now, I've talked about connections in space. But there are also connections in time. If the stuff that makes the universe is strongly connected in space and not usefully thought of as the aggregate sum of its parts, then attributing a cause of an event to a specific component may not be meaningful either. Just as you can't attribute the spin of a proton to any one of its constituents, you can't attribute an event in time to a single earlier cause. Complex systems have neither a useful notion of individuality nor a proper notion of causality.

THE BIGGER AN ANIMAL'S BRAIN, THE GREATER ITS INTELLIGENCE

NICHOLAS HUMPHREY

Psychologist; Darwin College, Cambridge; author,
Soul Dust: The Magic of Consciousness

The bigger an animal's brain, the greater its intelligence. You may think the connection is obvious. Just look at the evolutionary lineage of human beings. Humans have bigger brains, and are cleverer, than chimpanzees; and chimpanzees have bigger brains, and are cleverer, than monkeys. Or, as an analogy, look at the history of computing machines in the 20th century. The bigger the machines, the greater their number-crunching powers. In the 1970s, the new computer at my university department took up a whole room.

From the phrenology of the 19th century to the brain-scan sciences of the 21st, it has indeed been widely assumed that brain volume determines cognitive capacity. In particular, you'll find the idea repeated in every modern textbook—that the brain size of different primate species is causally related to their social intelligence. I admit I'm partly responsible for this, having championed the idea back in the 1970s. Yet for a good many years now, I've had a hunch that the idea is wrong.

There are too many awkward facts that don't fit in. For a start, we know that modern humans can be born with only two-thirds the normal volume of brain tissue and show next to no cognitive deficit as adults. We know that during normal human brain development, the brain actually shrinks as cognitive performance improves (a notable example being changes in the "social brain" during adolescence, where the cortical gray matter decreases in volume by about 15 percent between ages ten and twenty). And most surprising of all, we know that there are nonhuman animals, such as honeybees or parrots, that can emulate many feats of human intelligence with brains only a millionth (bee) or a thousandth (parrot) the size of a human's.

The key, of course, is programming: What really matters to cognitive performance isn't so much the brain's hardware as its onboard software. And smarter software doesn't need a bigger hardware base (in fact, as the shrinkage of the cortex during adolescence shows, it may actually need a smaller—tidier—one). It's true that programs to deliver superior performance may require a lot of designing, either by natural selection or learning. But once they've been invented, they'll likely make fewer demands on hardware than the older versions. To take the special case of social intelligence, I'd say it's quite possible that the algorithm for solving "theory of mind" problems could be written on the back of a postcard and implemented on an iPhone. In which case, the widely touted suggestion that the human brain had to double in size for humans to be capable of "second-order mind-reading" makes little sense.

Then why did the human brain double in size? Why is it much bigger than you might think it needs to be to underpin our level of intelligence? There's no question that big brains are costly to build and maintain. So if we're to retire the "obvious theory," what can we put in its place? The answer, I'd suggest, lies in the advantage of having a large amount of cognitive reserve. Big brains have spare capacity that can be called on if and when working parts get damaged or wear out. From adulthood onward, humans—like other mammals—begin to lose a significant amount of brain tissue to accidents, hemorrhages, and degeneration. But because humans can draw on this reserve, the loss doesn't have to show. This means humans can retain their mental powers into relative old age, long after their smaller-brained ancestors would have become incapacitated. (And as a matter of fact, the unfortunate individual born with an unusually small brain is much more likely to succumb to senile dementia in his forties.)

True, many of us die for other reasons, with unused brain power to spare. But some of us live considerably longer than we might have done if our brains were half the size. So, what evolutionary advantage does longevity bring, even the post-reproductive longevity typical of humans? The answer surely is that humans can benefit—as no other species could do—from the presence of mentally sound grandparents and great-grandparents, whose role in caretaking and teaching has been key to the success of human culture.

THE BIG BANG WAS THE FIRST MOMENT OF TIME

LEE SMOLIN

Physicist, Perimeter Institute, Waterloo, Ontario; author, Time Reborn

In my field of fundamental physics and cosmology, the idea most ready for retirement is that the Big Bang was the first moment of time.

In popular parlance, the Big Bang has two meanings. First, Big Bang cosmology is the hypothesis that our universe has been expanding for 13.8 billion years from an extremely hot and dense primordial state more extreme than the center of a star—or, indeed, than anywhere now existing. This I have no quarrel with; it's established scientific fact, which has been elaborated into a detailed story narrating the expansion of the universe from a uniform and dense hot plasma to the beautifully varied and complex world that is our home. We have detailed theories that pass numerous observational tests and explain the origins of all the structures we see, from the elements to galaxies, stars, planets, and the molecular building blocks of life. As in any good scientific theory, there are questions still to be answered, such as the precise nature of the dark matter and dark energy, which are prominent actors in the story, or the very interesting question of whether or not there was an early phase of inflationary exponential expansion. But these don't suggest that the basic picture is wrong.

What concerns me is the other meaning of Big Bang, which is the further hypothesis that the ultimate origin of our universe was a first moment of time, at which our universe was launched from a state of infinite density and temperature. According to this idea, nothing that exists is older than 13.8 billion years. It makes no sense to ask what was before that, because before that there wasn't even time.

The main problem with this second meaning of Big Bang is that it's not very successful as a scientific hypothesis, because it leaves big questions about the universe unanswered. It turns out that our universe had to have started in an extraordinarily special state to have evolved into anything like our universe. The hypothesis that there was a first moment of time is remarkably generic and unconstraining, as it's consistent with an infinite number of possible states in which the universe might have begun. This follows from a theorem proved by Stephen Hawking and Roger Penrose—that almost any expanding universe described by general relativity has such a first moment of time. Compared to almost all of these, our own early universe was extraordinarily homogeneous and symmetric. Why? If the Big Bang was the first moment of time, there can be no scientific answer, because there was no "before" on which to base an explanation. At this point, theologians see their opening and, indeed, have been lining up at the gates of science to impose their kind of explanation: that God made the universe and made it so.

Similarly, if the Big Bang was the first moment of time, there can be no scientific answer to the question of what chose the laws of nature. This leaves the field open to explanations such as the anthropic multiverse, which are unscientific because they call on unobservable collections of other universes and make no predictions by which their hypotheses might be tested and falsified.

There is, however, a chance for science to answer these

questions, and that's if the Big Bang was *not* the first moment of time but a transition from an earlier era of the universe—an era that *can* be investigated scientifically, because processes acting then would have given rise to our world.

For there to have been a time before the Big Bang, the Hawking-Penrose theorem must fail. But there's a simple reason to think it must: General relativity is incomplete as a description of nature, because it leaves out quantum phenomena. Unifying quantum physics with general relativity has been a major challenge for fundamental physics, one on which there has been much progress in the last thirty years. In spite of the absence of a definitive solution to the problem, there's robust evidence from quantum cosmology models that the infinite singularities forcing time to stop in general relativity are eliminated, turning the Big Bang—in the sense of a first moment of time—into a Big Bounce, which allows time to continue to exist before the Big Bang, deep into the past. Detailed models of quantum universes show a prior era ending with a collapse, where the density increases to very high values, but before the universe becomes infinitely dense, quantum processes take over which bounce the collapse into an expansion, launching a new era that could be our expanding universe.

There are currently several scenarios under study for what happened in the era before the Big Bang and how it transitioned to our expanding universe. Two of them hypothesize a quantum bounce and go under the name of "loop quantum cosmology" and "geometrogenesis." Two others—due respectively to Roger Penrose and to Paul Steinhardt and Neil Turok—describe cyclic scenarios in which universes die giving rise to new universes. A fifth posits that new universes are launched when quantum effects bounce black-hole singularities. These scenarios offer insights as to how the laws of nature governing our universe might have been chosen, and they may also explain how the initial conditions of our universe evolved from the previous universe. The important thing is that each of these hypotheses makes predictions for real, doable observations by which they might be falsified and distinguished from one another.

During the 20th century, we learned a great deal about the first three minutes (in Steven Weinberg's phrase) of our expanding universe. During this century, we can look forward to gaining scientific evidence of the last three minutes of the era before ours and learning how physics before the Big Bang gave rise to the birth of our universe.

THE UNIVERSE BEGAN IN A STATE OF EXTRAORDINARILY LOW ENTROPY

ALAN GUTH

Cosmologist; Victor F. Weisskopf Professor of Physics, MIT; inaugural winner, Milner Foundation's Fundamental Physics Prize; author, The Inflationary Universe

The roots of this issue go back at least to 1865, when Rudolf Clausius coined the term "entropy" and stated that the entropy of the universe tends to a maximum. This idea is now known as the second law of thermodynamics, which is most often stated thus: The entropy of an isolated system always increases or stays constant but never decreases. Isolated systems tend to evolve toward the state of maximum entropy—the state of thermodynamic equilibrium. Even though entropy will play a crucial role in this discussion, it will suffice to use a fairly crude definition: Entropy is a measure of the disorder of a physical system. In the underlying quantum description, entropy is a measure of the number of quantum states corresponding to a given description in terms of macroscopic variables such as temperature, volume, and density.

The classic example is a gas in a closed box. If we start with all the gas molecules in a corner of the box, we can imagine watching what happens next. The gas molecules will fill the box, increasing the entropy to the maximum. But it never goes the other way: If the gas molecules fill the box, we'll never see them spontaneously collect into one corner.

This behavior seems natural but is hard to reconcile with

our understanding of the underlying laws of physics. The gas makes a huge distinction between the past and the future, always evolving toward larger entropy in the future. This one-way behavior of matter in bulk is called the "arrow of time." Nonetheless, the microscopic laws that describe collisions of molecules are time-symmetric, making no distinction between past and future.

Any movie of a collision could be played backward and it would also show a valid picture of the collision. (To account for some very rare events discovered by particle physicists, the movie is guaranteed to be valid only if it is also reflected in a mirror and has every particle relabeled as the corresponding antiparticle. But these complications don't change the key issue.) There's an important problem, therefore, which is over a century old: to understand how the arrow of time could possibly arise from time-symmetric laws of evolution.

The arrow-of-time mystery has driven physicists to seek possible causes within the laws of physics we observe, but in vain. The laws make no distinction between the past and the future. Physicists have understood, however, that a low entropy state is always likely to evolve into a higher entropy state simply because there are many more states of higher entropy. Thus, the entropy today is higher than the entropy yesterday, because yesterday the universe was in a low entropy state. And it was in a low entropy state yesterday because the day before that it was in an even lower entropy state. The traditional understanding follows this pattern back to the origin of the universe, attributing the arrow of time to some not-well-understood property of cosmic initial conditions, which created the universe in a special low entropy state. As Brian Greene wrote in *The Fabric of the Cosmos*:

> The ultimate source of order, of low entropy, must be the big bang itself. . . . The egg splatters rather than unsplatters because it is carrying forward the drive toward higher entropy that was initiated by the extraordinarily low entropy state with which the universe began.

Based on an elaboration of a 2004 proposal by Sean Carroll and Jennifer Chen, there's a possibility of a new solution to the age-old problem of the arrow of time. This work, by Sean Carroll, Chien-Yao Tseng, and me, is still in the realm of speculation and hasn't yet been vetted by the scientific community. But it seems to provide an attractive alternative to the standard picture.

The standard picture holds that the initial conditions for the universe must have produced a special low-entropy state because one is needed to explain the arrow of time. (No such assumption is applied to the final state, so the arrow of time is introduced through a time-asymmetric condition.) We argue, to the contrary, that the arrow of time can be explained without assuming a special initial state, so there's no longer any motivation for the hypothesis that the universe began in a state of extraordinarily low entropy. The most attractive feature of this idea is that there's no longer a need to introduce *any* assumptions that violate the time symmetry of the known laws of physics.

The basic idea is simple: We don't really know if the maximum possible entropy for the universe is finite or infinite, so let's assume it's infinite. Then, no matter what entropy the universe started with, the entropy would have been low compared with its maximum. That's all that's needed to explain why the entropy has been rising ever since! The metaphor of gas in a box is replaced by a gas with no box. In the context of what physicists call a toy model—that is, one meant to illustrate a basic principle without trying to be otherwise realistic—we can imagine choosing, in a random and time-symmetric way, an initial state for a gas composed of some finite number of noninteracting particles. It's important here that any well-defined state will have a finite value for the entropy and a finite value for the maximum distance of any particle from the origin of our coordinate system. If such a system is followed into the future, the particles might move inward or outward for some finite time, but ultimately the inward-moving particles will pass the central region and will start moving outward. All particles will eventually be moving outward, and the gas will continue indefinitely to expand into the infinite space, with the entropy rising without limit. An arrow of time—the steady growth of entropy with time—has been generated, without introducing any time-asymmetric assumptions.

An interesting feature of this picture is that the universe need not have a beginning but could be continued from where we started in both directions of time. Since the laws of evolution and the initial state are time-symmetric, the past will be statistically equivalent to the future. Observers in the deep past will see the arrow of time in the opposite direction from ours, but their experience will be no different from ours.

ENTROPY

BRUCE PARKER

Oceanographer; visiting professor, Center for Maritime Systems, Stevens Institute of Technology; author, The Power of the Sea

Could one really have the nerve to suggest "retiring" the idea of entropy? I don't believe we abandon old ideas before new ones are developed. Old ideas disappear, or are modified, only when new, better ideas are developed; they're never just retired. So, no, we shouldn't retire entropy, but perhaps we should treat it with a little less importance and recognize the paradox it creates.

Entropy, the measure of the degree of disorder in a system, has held a lofty place in physics—being part of a law, no less, not just a theory. The second law of thermodynamics says that in any closed system, entropy always increases with time. Unless work is done to prevent it, a closed system will eventually reach maximum entropy and a state of thermal equilibrium. Max Planck believed that entropy (along with energy) was the most important property of physical systems. In *The Nature of the Physical World* (1927), Sir Arthur Eddington wrote: "The law that entropy always increases—the second law of thermodynamics—holds, I think, the supreme position among the laws of Nature." But as a young physics student in college, I must admit I never understood their excitement (and I was not the only student to be unimpressed). The second law seemed of minor importance compared with the first law of thermodynamics, the conservation of energy: Energy could be transformed into different forms, but it was always conserved. The first law had beautiful partial differential equations (as do all the conservation equations of physics), whose solutions accurately described and predicted so much of the world and literally changed our lives. The second law was not a conservation equation and had no beautiful partial differential equations. It wasn't even an equality. Has the idea of entropy and the second law had any major effect on science and engineering or changed the world?

The second law was a statistical law, initially a generalization of conclusions reached when looking at the motion of molecules or particles. As students, we could easily understand the classic example of how hot (fast-moving) molecules on one side of a closed box mixed with cold (slowing-moving) molecules on the other side, and why they couldn't separate again once they were together and all at the same temperature. We understood why it was irreversible. And we understood the concept of the "arrow of time." Sure, the mathematics of the first law (and the other conservation equations of physics) worked in both directions of time—but with initial conditions and boundary conditions, we always knew which way things moved. It didn't seem to require another law. In fact, the second law (as applied now to all situations) seemed to be an assumption rather than a law. Especially when it was applied to the entire universe, which we understand so little about.

Where the universe (whatever that entails, which may be more than our presently visible/observable universe) is concerned, the first law tells us that all the energy in it will be conserved, although perhaps converted into various forms. But the second law says that at some time in the future no more energy transformations can take place. The universe will reach some stage of maximum entropy and thermal equilibrium.

The second law essentially says that the universe must have had a beginning and an end. That's very difficult to accept. The universe must be timeless, for if there was a beginning, what was there before this beginning? Something cannot come out of nothing (and by "nothing," I mean the lack of anything, even things we don't know about yet).

Of course, the present Big Bang theory posits a beginning (of sorts), and our present form of the universe has apparently expanded from a singularity, but we don't know what came before that, and oscillating models of the universe are being proposed so that the universe is timeless. With such models, if entropy is very high at the end of our universe and was very low at the beginning of our universe, what process could essentially reset entropy to a low value? As relates to an oscillating universe, should entropy perhaps be conserved somehow? Could there be some type of energy conversion that doesn't require work (in our classical sense)? Could the universe actually be the one and only possible perpetual-motion machine (a contraption forbidden by the second law)? If existence is endless in time, it would seem so.

The whole idea of entropy has always felt wrong or misplaced in other ways also. We talk about the universe going from order to disorder. Yet this supposed order is merely the compression of all the universe's matter in some tiny volume—a singularity—and when it expanded there was less order because the particles were more spread out. And yet order is being created all the time.

The greatest result of our expanding and evolving universe is its ever increasing complexity: first, from gravity condensing matter, then from supernovae creating higher-number elements, then from chemical evolution, and then from biological evolution driven by natural selection, which culminated in self-reproducing life and eventually the incredible complexity of our brains.

Complexity is synonymous with low entropy. The expanding universe has countless small (relative to the universe's size) pockets of extremely low entropy surrounded by vast areas of higher entropy (much of which resulted from the creation of these low-entropy areas). Are the higher orders of complexity (and thus lower orders of entropy) taken into account in attempts to balance the entropy of the universe? Many current scientific papers in cosmology try to sum up the universe's total entropy with formulas that may well be far too simple to account for all the as-yet-unknown physics going on in our strange universe.

We cannot retire entropy, but should we maybe rethink it?

THE UNIFORMITY AND UNIQUENESS OF THE UNIVERSE

ANDREI LINDE

Theoretical physicist, Stanford University; father of eternal chaotic inflation; inaugural winner, Milner Foundation's Fundamental Physics Prize, 2012

For most of the 20th century, scientific thought was dominated by the idea of the uniformity of the universe and the uniqueness of the laws of physics. Indeed, cosmological observations indicated that the universe on the largest possible scales is almost exactly uniform, with an accuracy better than 1 in 10,000.

The situation is similar with respect to the uniqueness of the laws of physics. We knew, for example, that the electron mass is the same everywhere in the observable part of the universe, so the obvious assumption was that it took the same value everywhere—that it was just a constant of nature. For a long time, one of the great goals of physics was to find a single theory—a Theory of Everything—that would unify all fundamental interactions and provide an unambiguous explanation for all known parameters of particle physics.

Some thirty years ago, a possible explanation arose for the uniformity of the universe. The main idea was that our part of the world emerged as a result of an exponentially rapid stretching of space called cosmic inflation. As all "wrinkles" and nonuniformities of space stretched out and disappeared, the universe became incredibly smooth. Add some quantum fluctuations, stretch them, and the uniformity became just a little bit less perfect, and galaxies emerged.

At first, inflationary theory looked like the exotic product of a vivid imagination. But thanks to the enthusiastic work of thousands of scientists, its various predictions have been confirmed by many observations made by COBE, WMAP, *Planck*, and most recently by BICEP2. If, as I think, the theory is correct, we finally have a scientific explanation of why the world is so uniform.

But inflation does not predict that this uniformity must extend beyond the observable part of the universe. To give an analogy: Suppose the universe is the surface of a big soccer ball consisting of black and white hexagons. If we inflate it, the size of each white or black part becomes exponentially large. If inflation is powerful enough, those who live in a black part of the universe will never see a white part. They will believe the whole universe is black, and they will try to scientifically explain why it cannot be any other color. Those who live in a white universe will never see the black parts and will therefore think the whole world must be white. But black and white parts alike can coexist in an inflationary universe without contradicting observations.

Unlike the black/white analogy, in physics the number of different "colors"—that is, different states of matter—can be exponentially large. The best current candidate for a Theory of Everything is string theory, which can be successfully formulated in spacetime with ten dimensions (nine of space and one of time). But we live in a universe with three dimensions of space. Where are the six others? The answer is that they're compactified—squeezed so small that we cannot move in those directions—and so we perceive the world as if it were three-dimensional.

From the early days of string theory, physicists knew there

were exponentially many different ways to compactify the extra six dimensions, but we didn't know what kept the compactified dimensions from expanding. This problem was solved about ten years ago, and the solution validated the earlier expectation of the exponentially large number of possibilities. Some estimates are as large as 10^{500}. And each of these options describes a part of the universe with a different vacuum energy and different types of matter. In the context of the inflationary theory, this means the world may well consist of 10^{500} huge universes with different types of matter.

A pessimist would argue that since we don't see other parts of the universe, we cannot prove that this picture is correct. An optimist will counter that neither can we disprove it, because its main assumption is that the other universes are far away from us. And since we know that our best current theory allows 10^{500} different universes, anybody who argues that the universe must have the same properties everywhere has to prove that only one of those 10^{500} universes is possible.

And there's something else: There are many strange coincidences in our world. The mass of the electron is 2,000 times smaller than the mass of the proton. Why? The only "reason" is that if it were even a little different, life as we know it would be impossible. The masses of the proton and neutron almost coincide. Why? If the mass of either were even a little different, life as we know it would be impossible. The energy of empty space in our part of the universe is not zero but a tiny number—more than a 100 orders of magnitude below the naïve theoretical expectations. Why? The only explanation we have is that we couldn't live in a world with a larger vacuum energy.

The correlation between our properties and the properties of the world is called the anthropic principle. But if the universe came in only one copy, this correlation would not explain why. We'd need to speculate about a divine cause making the universe custom-built for humans. However, with a multiverse consisting of many different parts with different properties, the correlation between our properties and the properties of the part of the world we live in makes perfect sense.

Can we return to the old picture of a single universe? Possibly. But in order to do so, we must (1) invent a better cosmological theory, (2) invent a better theory of fundamental interactions, and (3) propose an alternative explanation for the aforementioned miraculous coincidences.

INFINITY

MAX TEGMARK

Physicist, cosmologist, MIT; scientific director, Foundational Questions Institute; author, Our Mathematical Universe

I was seduced by infinity at an early age. Georg Cantor's diagonality proof that some infinities are bigger than others mesmerized me, and his infinite hierarchy of infinities blew my mind. The assumption that something truly infinite exists in nature underlies every physics course I've ever taught at MIT—and, indeed, all of modern physics. But it's an untested assumption, which begs the question: Is it actually true?

There are in fact two separate assumptions: "infinitely big" and "infinitely small." By infinitely big, I mean that space can have infinite volume, that time can continue forever, and that there can be infinitely many physical objects. By infinitely small, I mean the continuum—the idea that even a liter of space contains an infinite number of points, that space can be stretched out indefinitely without anything bad happening, and that there are quantities in nature that can vary continuously. The two assumptions are closely related, because inflation, the most popular explanation of our Big Bang, can create an infinite volume by stretching continuous space indefinitely.

The theory of inflation has been spectacularly successful and is a leading contender for a Nobel Prize. It explains how a subatomic speck of matter transformed into a massive Big Bang, creating a huge, flat, uniform universe, with tiny density fluctuations that eventually grew into today's galaxies and cosmic large-scale structure—all in beautiful agreement with precision measurements from experiments such as the *Planck* and the BICEP2 experiments. But by predicting that space isn't just big but truly infinite, inflation has also brought about the so-called measure problem, which I view as the greatest crisis facing modern physics. Physics is all about predicting the future from the past, but inflation seems to sabotage this. When we try to predict the probability that something particular will happen, inflation always gives the same useless answer: infinity divided by infinity. The problem is that whatever experiment you make, inflation predicts there will be infinitely many copies of you, far away in our infinite space, obtaining each physically possible outcome; and despite years of teeth-grinding in the cosmology community, no consensus has emerged on how to extract sensible answers from these infinities. So, strictly speaking, we physicists can no longer predict anything at all!

This means that today's best theories need a major shakeup by retiring an incorrect assumption. Which one? Here's my prime suspect: ∞.

A rubber band can't be stretched indefinitely, because although it seems smooth and continuous, that's merely a convenient approximation. It's really made of atoms, and if you stretch it too far, it snaps. If we similarly retire the idea that space itself is an infinitely stretchy continuum, then a big snap of sorts stops inflation from producing an infinitely big space and the measure problem goes away. Without the infinitely small, inflation can't make the infinitely big, so you get rid of both infinities in one fell swoop—together with many other problems plaguing modern physics, such as infinitely dense black-hole singularities and infinities popping up when we try to quantize gravity.

In the past, many venerable mathematicians were skeptical of infinity and the continuum. The legendary Carl Friedrich Gauss denied that anything infinite really exists, saying "Infinity is merely a way of speaking" and "I protest against the use of infinite magnitude as something completed, which is never permissible in mathematics." In the past century, however, infinity has become mathematically mainstream, and most physicists and mathematicians have become so enamored with infinity that they rarely question it. Why? Basically, because infinity is an extremely convenient approximation for which we haven't discovered convenient alternatives.

Consider, for example, the air in front of you. Keeping track of the positions and speeds of octillions of atoms would be hopelessly complicated. But if you ignore the fact that air is made of atoms and instead approximate it as a continuum—a smooth substance that has a density, pressure, and velocity at each point—you'll find that this idealized air obeys a beautifully simple equation explaining almost everything we care about: how to build airplanes, how we hear them with sound waves, how to make weather forecasts, and so forth. Yet despite all that convenience, air of course isn't truly continuous. I think it's the same way for space, time, and all the other building blocks of our physical word.

Let's face it: Despite their seductive allure, we have no direct observational evidence for either the infinitely big or the infinitely small. We speak of infinite volumes with infinitely many planets, but our observable universe contains only about 10^{89} objects (mostly photons). If space is a true continuum, then to describe even something as simple as the distance between two points requires an infinite amount of information, specified by a number with infinitely many decimal places. In practice, we physicists have never managed to measure anything to more than about seventeen decimal places. Yet real numbers, with their infinitely many decimals, have infested almost every nook and cranny of physics, from the strengths of electromagnetic fields to the wave functions of quantum mechanics. We describe even a single bit of quantum information (qubit) using two real numbers involving infinitely many decimals.

Not only do we lack evidence for the infinite but we don't need the infinite to do physics. Our best computer simulations, accurately describing everything from the formation of galaxies to tomorrow's weather to the masses of elementary particles, use only finite computer resources by treating everything as finite. So if we can do without infinity to figure out what happens next, surely nature can, too—in a way that's more deep and elegant than the hacks we use for our computer simulations. Our challenge as physicists is to discover this elegant way and the infinity-free equations describing it—the true laws of physics. To start this search in earnest, we need to question infinity. I'm betting that we also need to let go of it.

THE LAWS OF PHYSICS ARE PREDETERMINED

LAWRENCE M. KRAUSS

Physicist, cosmologist; director, Origins Project, Arizona State University; author, A Universe from Nothing

Einstein once said, "What really interests me is whether God had any choice in the creation of the universe." By "God," of course, he didn't mean God. What he was referring to was the question that has driven most scientists who, like me, are attempting to unravel the fundamental laws governing the cosmos at its most basic scale: Is there only one consistent set of physical laws? If we change one fundamental constant, one force law, would the whole edifice tumble?

Most scientists of my generation, like Einstein before us, implicitly assumed that the answer to these questions was yes. We wanted to uncover the One True Theory, the mathematical formulation explaining why there had to be four forces in nature, why the proton is 2,000 times heavier than the electron, and so on. In recent memory, this effort reached its most audacious level in the 1980s, when string theorists argued that they had found the Theory of Everything—that using the postulates of string theory, one would be driven to a unique physical theory, with no wiggle room, that would ultimately explain at a fundamental level everything we see.

Needless to say, that grand notion has had to be put aside for now, as string theory has failed—thus far, at least—to live up to such lofty promises. In the process, however, in part driven by string theory's lack of success, we've embraced the opposite alternative: The laws of nature we measure may be totally accidental, local to our environment (namely, our universe), not prescribed with robustness by any universal principle, and by no means generic or required.

String theory, for example, suggests a host of new possible dimensions. To make contact with our observed four-dimensional universe, it requires these other dimensions to be invisible, by curling up on such small scales that they cannot be probed; or it requires the known forces and particles to be restricted to a four-dimensional "brane." But it appears that there are many, many different ways to hide the extra dimensions, and each one produces a different four-dimensional universe with different laws. It also appears that four dimensions themselves need not be universal; perhaps there are two-dimensional universes, or six-dimensional ones.

You don't have to go to such speculative heights to reach the possible conclusion that the laws of our universe may have come into existence when our universe did. The theory of inflation, currently the best theory for how our universe obtained the characteristics it's measured to have on large scales, suggests that at very early times there was a runaway period of expansion. In different places, and perhaps different times, small regions will stop "inflating," as a cosmic phase transition occurs in those regions, changing the stable configuration of particles and fields. But in this picture, most of the "metaverse," if you will, is still inflating, and each region that departs from inflation, each universe, can settle into a different state with different laws, just as ice crystals on a window can form in different directions.

All this strongly suggests there may be nothing fundamental whatsoever about the "fundamental" laws we measure in our universe. They could simply be accidental. Physics becomes, in this sense, an environmental science.

Now, many people have picked up on this notion to suggest that somehow we can understand our laws because they're selected anthropically—that is, if they were any different, life wouldn't have developed in our universe. However, this idea is full of problems—not least because we have no idea what possibilities exist, and whether changing a few or a huge number of fundamental parameters could result in viable, habitable universes. We also have no idea whether or not we're typical life-forms. Most life that evolves or will evolve in our universe might be quite different.

Focusing on anthropics misses the point in any case. The important fact is that we must be willing to give up the idea that the laws of physics in our universe reflect some underlying fundamental order, that the laws are somehow preordained by principles of beauty or symmetry. There's nothing new about this. It was myopic to assume that life on our planet was preordained. We now understand that accidents of natural selection and environmental traumas governed the history of life that led to our existence. It's equally myopic to assume we are somehow the pinnacles of evolution—that all roads lead to us, or that we won't lead to something completely different in the future.

It's myopic to assume that the universe we now live in will always be this way. It won't be. As several of us have argued, it seems that in the far future all the galaxies we now see will disappear. But it may be much worse. It's myopic to assume our laws are universal in time and space, even in our universe. Current data on the Higgs particle suggests that the universe could yet again undergo a cosmic phase transition that would change the stable forces and particles, and we and everything we see might disappear.

We've come to accept the notion that life is not preordained; we also need to give up the quaint notion that the laws of physics *are*. Cosmic accidents are everywhere—it's possible that our entire universe is just another accident.

THEORIES OF ANYTHING

PAUL STEINHARDT

Albert Einstein Professor in Science, Departments of Physics and Astrophysical Sciences, Princeton University; coauthor (with Neil Turok), Endless Universe

A pervasive idea in fundamental physics and cosmology that should be retired: the notion that we live in a multiverse in which the laws of physics and the properties of the cosmos vary randomly from one patch of space to another. According to this view, the laws and properties within our observable universe cannot be explained or predicted, because they're set by chance. Different regions of space too distant to ever be observed have different laws and properties, according to this picture. Over the entire multiverse there are infinitely many distinct patches. Among these patches, in the words of Alan Guth, "anything that can happen will happen; in fact, it will happen an infinite number of times."[a] Hence, I refer to this concept as a Theory of Anything.

Any observation or combination of observations is consistent with a Theory of Anything. No observation or combination of observations can disprove it. Proponents seem to revel in the fact that the theory cannot be falsified. The rest of the scientific community should be up in arms, since an unfalsifiable idea lies beyond the bounds of normal science. Yet, except for a few voices, there has been surprising complacency about (and in some cases grudging acceptance of) a Theory of Anything as a logical possibility. The scientific journals are full of papers treating the Theory of Anything seriously. What's going on?

Have experiments revealed that our observable universe and the fundamental laws are too complicated to be explained by normal science? Absolutely not! Quite the opposite. On the macroscopic scale, the latest measurements show our observable universe to be remarkably simple, described by a very few parameters, obeying the same physical laws throughout, and exhibiting remarkably uniform structure in all directions. On the microscopic scale, the Large Hadron Collider at CERN has revealed the existence of the Higgs boson, in accord with what theorists predicted nearly fifty years ago based on sound scientific reasoning.

A simple outcome calls for a simple explanation for why it had to be so. Why, then, consider a Theory of Anything that allows any possibility, including complicated ones? The motivation is the failure of two favorite theoretical ideas—inflationary cosmology and string theory. Both were thought to produce a unique outcome. Inflationary cosmology was invented to transform the entire cosmos into a smooth universe populated by a scale-invariant distribution of hot spots and cold spots, just as we observe it to be. String theory was supposed to explain why elementary particles could have only the precise masses and forces they do. After more than thirty years' investment in each of these ideas, theorists have found them unable to achieve those ambitious goals. Inflation, once started, runs eternally and produces a multiverse of "pockets," whose properties vary over every conceivable possibility—flat and nonflat, smooth and nonsmooth, scale-invariant and not scale-invariant, and so on. Despite laudable efforts by many theorists to save the theory, there's no solid reason known today why inflation should cause our observable universe to be in a pocket with the smoothness and other very simple properties we observe. A continuum of other conditions is equally possible.

In string theory, a similar explosion of possibilities has occurred, driven by attempts to explain the 1998 discovery of the accelerated expansion of the universe. The acceleration is thought to be due to positive vacuum energy, an energy associated with empty space. Instead of predicting a unique possibility for the vacuum state of the universe and particles and fields that inhabit it, our current understanding of string theory is that there's a complex landscape of vacuum states corresponding to exponentially different kinds of particles and different physical laws. The set of vacuum space contains so many possibilities that surely, it is claimed, one will include the right amount of vacuum energy and the right kinds of particles and fields. Mix inflation and string theory and the unpredictability multiplies. Now every combination of macrophysical and microphysical possibilities can occur.

I suspect the theories would never have gained the acceptance they have if these problems had been broadly recognized at the outset. Historically, if a theory failed to achieve its goals, it was improved or retired. In this case, though, the commitment to the theories has become so strong that some prominent proponents have seriously advocated moving the goalposts. They say we should be prepared to abandon the old-fashioned idea that scientific theories should give definite predictions and accept a Theory of Anything as the best that can ever be achieved.

I draw the line there. Science is useful insofar as it explains and predicts why things are the way they are and not some other way. The worth of a scientific theory is gauged by the number of do-or-die experimental tests it passes. A Theory of

Anything is useless, because it doesn't rule out any possibility, and worthless because it submits to no do-or-die tests. (Many papers discuss potential observable consequences, but these are only possibilities, not certainties, so the Theory of Anything is never really put at risk.)

A priority for theorists today is to determine if inflation and string theory can be saved from devolving into a Theory of Anything—and if not, to seek new ideas to replace them. Because an unfalsifiable Theory of Anything creates unfair competition for real scientific theories, leaders in the field can play an important role by speaking out—making it clear that Anything is not acceptable—to encourage talented young scientists to rise to the challenge. The sooner we can retire the Theory of Anything, the sooner this important science can progress.

M-THEORY/STRING THEORY IS THE ONLY GAME IN TOWN

ERIC R. WEINSTEIN
Mathematician and economist; managing director, Thiel Capital

If one views science as an economist, it stands to reason that the first scientific theory to be retired would be the one offering the greatest opportunity for arbitrage in the marketplace of ideas. Thus it's not enough to look for ideas that are merely wrong; we should look instead for troubled scientific ideas that block progress by inspiring zeal, devotion, and what biologists politely term "interference competition" all out of proportion to their history of achievement. Here it's hard to find a better candidate for an intellectual bubble than that which has formed around the quest for a consistent Theory of Everything physical, reinterpreted as if it were synonymous with "quantum gravity." If Nature were trying to send a polite message that there's other work to be done before we quantize gravity, it's hard to see how she could send a clearer message than dashing the Nobel aspirations of two successive generations of Niels Bohr's brilliant descendants.

To recall: Modern physics rests on a stool with three classical geometric legs, fashioned individually by Albert Einstein, James Clerk Maxwell, and P. A. M. Dirac. The last two of those legs can be together retrofitted to a quantum theory of force and matter known as the Standard Model, while the first stubbornly resists any such attempt at an upgrade, rendering the semi-quantum stool unstable and useless. It's from this that the children of Bohr have derived the need to convert the children of Einstein to the quantum religion at all costs, so that the stool can balance.

But, to be fair to those who insist that Einstein must now be made to bow to Bohr, the most strident of those enthusiasts have offered a reasonable challenge. Quantum exceptionalists claim, despite an unparalleled history of non-success, that string theory (now rebranded as M-theory, variously standing for "matrix," "membrane," "mother," or "magic") remains the only game in town, because fundamental physics has gotten so hard that no one can think of a credible alternate unification program. If we're to dispel this as a canard, we must make a good-faith effort to answer the challenge by providing interesting alternatives, lest we be left with nothing at all.

My reason for believing there's a better route to the truth is that out of what seems misplaced loyalty to our beloved Einstein, we've been too reverential to the exact form of general relativity. For example, if before retrofitting we look closely at the curvature and geometry of the three legs of the stool, we can see something striking: They're subtly incompatible at a classical geometric level before any notion of a quantum is introduced. Einstein's leg seems the sparest and sturdiest, clearly showing the attention to function found in the school of "intrinsic geometry" founded by the German Bernhard Riemann. The Maxwell and Dirac legs are somewhat more ornamented, exploring the freedom of form that's the *raison d'être* for a more whimsical school of "auxiliary geometry" pioneered by the Alsatian Charles Ehresmann. This leads us naturally to a very different question: What if the quantum incompatibility of the existing theories is really a red herring with respect to unification, and the real sticking point is a geometric conflict between the mathematicians Ehresmann and Riemann rather than an incompatibility between the physicists Einstein and Bohr? Even worse, it could be that none of the foundations are ready to be quantized. What if all three theories are subtly incomplete at a geometric level and the quantum will follow only when all three are retired and replaced with a unified geometry?

If such an answer exists, it cannot be expected to be a generic geometric theory, because all three of the existing theories are in some sense the simplest possible in their respective domains. Such a unified approach might instead involve a new kind of mathematical toolkit, combining elements of the two major geometric schools, which would be relevant to physics only if the observed world can be shown to be of a very particular subtype. Happily, with the discoveries of neutrino mass, nontrivial dark energy, and dark matter, the world we see looks increasingly to be of the special class that could accommodate such a hybrid theory.

I could go on in this way, but it isn't the only interesting line of thinking. Whereas ultimately there may be a single unified theory to summit, there are few such intellectual peaks that can be climbed from just one face. We need to return physics to its natural state of individualism, so that independent researchers need not fear large research communities that, in the quest for mindshare and resources, crowd out isolated rivals pursuing genuinely interesting inchoate ideas that head in new directions. Unfortunately, it's difficult to responsibly encourage theorists who aren't independently wealthy to develop truly speculative theories in a community that now applies artificially strict standards to new programs and new voices while letting M-theory stand, year after year, for "mulligan" and *mañana.*

Established string theorists may, with a twinkle in the eye, shout, "Predictions!," "Falsifiability!," or "Peer review!" at younger competitors in jest. Yet potentially rival "infant industry" research programs, as the saying goes, do not die in jest but in earnest. Given the history of scientific exceptionalism surrounding quantum gravity research, it's neither desirable nor necessary to retire M-theory explicitly, as it contains many fascinating ideas. Instead, we need only insist that the training wheels once customarily circulated to new entrants to reinvigorate the community be transferred to emerging candidates from those who have monopolized them for decades. We can then wait, at long last, to see whether "the only game in town," when denied the luxury of special pleading by senior boosters, has the support from Nature to stay upright.

STRING THEORY

FRANK TIPLER

Professor of mathematical physics, Tulane University; coauthor (with John Barrow), The Anthropic Cosmological Principle; *author,* The Physics of Immortality

In his *Scientific Autobiography*, Max Planck recalls being unable to persuade the chemist Wilhelm Ostwald that the second law of thermodynamics could not be deduced from the first law of thermodynamics. "This experience gave me also an opportunity to learn a fact—a remarkable one, in my opinion: A new scientific truth does not triumph by convincing its opponents and making them see the light, but rather because its opponents eventually die, and a new generation grows up that is familiar with it." Planck also wrote of his conflict with Ostwald:

> It is one of the most painful experiences of my entire scientific life that I have but seldom—in fact, I might say, never—succeeded in gaining universal recognition for a new result, the truth of which I could demonstrate by a conclusive, albeit only theoretical proof. This is what happened this time, too. All my sound arguments fell on deaf ears. It was simply impossible to be heard against the authority of men like Ostwald, Helm and Mach.[b]

Fortunately, Planck was able to obtain universal recognition for his radiation law—again, not by his theoretical proof but by experimental confirmation.

There has been a tendency among theoretical physicists, particularly string theorists, to downplay the importance of experimental confirmation in recent years. Many have even claimed that Copernicus was not superior in predictive power over Ptolemy. I myself decided to check this claim, by looking at Tycho's notebooks. I discovered that between 1564 and 1601, Tycho compared Copernicus' predictions and Ptolemy's predictions with his own observations 294 times. As I expected, Copernicus was superior. So Copernicus' theory was confirmed as experimentally superior to Ptolemy's long before Galileo. So I have put the Copernicus-was-no-better-than-Ptolemy idea to the (historical) experimental test and found it false: Copernicus trumps Ptolemy.

As it was in the beginning of modern science, so it should be now. Experimental confirmation is the hallmark of true science. Since string theorists have failed to propose *any* way to confirm string theory experimentally, string theory should be retired, today, now.

OUR WORLD HAS ONLY THREE SPACE DIMENSIONS

GORDON KANE

Victor F. Weisskopf Distinguished University Professor of Physics, University of Michigan; author, Supersymmetry and Beyond

Physics theories typically predict aspects of the world which we don't see. For example, Maxwell's theory of electromagnetism correctly predicted that the spectrum of light we see was just a part of the full spectrum, which extends into infrared and ultraviolet waves invisible to us. String theory predicts that our world has more than three space dimensions. Contrary to much that is written and said, string theory is broadly predictive and testable. Before I explain its testability, I'll describe why great progress in making a comprehensive underlying theory of our physical world may emerge from formulating theories in more than three space dimensions. I'll call it a "final theory," following Steven Weinberg.

What could we gain by giving up the idea that our world has three space dimensions? String theory emerged when John Schwarz and Michael Green noticed in the summer of 1984 that it was possible to write a mathematically consistent quantum theory of gravity in ten dimensions. That's a big gain and clue. For me and some other theorists, it's even more important that string theories address all, or nearly all, of the issues and questions that need answering in order to have a final theory. There's been major progress here in the past decade. The initial excess optimism of string theorists caused an overcompensation, now tempered by increasingly many results. The highly successful and well-tested four-dimensional so-called Standard Models of particle physics and of cosmology provide powerful, accurate, and complete (with the discovery of the Higgs boson) descriptions of the world we see, but they don't provide explanations and understanding for a number of issues that *are* addressed by string theory. The success of the Standard Model(s) is strong evidence that sticking with the four-dimensional world gets in the way of going beyond description to explaining and understanding.

To explain our universe, obviously the higher-dimension string theories have to be projected onto a four-dimensional universe, a process with the understandable but unfortunate name "compactification" (for historical reasons). Experiments and observations have to be done in our 4-D universe, so only compactified theories can be directly tested. Compactified string theories address why the universe is mainly made of matter and not antimatter, what the dark matter is, why quarks and leptons come in three similar families, what the individual quark and lepton masses are, the existence of the Higgs mechanism and how it gives mass to quarks and leptons and force-carrying bosons, cosmological history from the end of inflation to the origins of nuclei (after which the Standard Model takes over), the cause of inflation, and much more. Compactified string theories successfully predicted (before the measurements) the mass and properties of the Higgs boson discovered at CERN in 2012 and make predictions for the existence of "supersymmetric partner particles," some of which should be produced and detected at the upgraded CERN collider in 2015, if it functions as planned. Examples already exist in compactified string theories for all of these. This is research in progress, so much still needs to be worked out and understood better—and tested at colliders and in dark-matter and other experiments—but we can already see that all these exciting opportunities exist.

In 1995, Edward Witten argued that there was an eleven-dimensional theory he called M-theory which could give a consistent quantum theory of gravity and could be projected onto several 10-D string theories in different ways. They had names like heterotic or Type II. Those 10-D theories could then be compactified to 4-D theories (with six small, curled-up dimensions) and make testable predictions as described above. M-theory can also be compactified directly onto a 7-D curled-up (G_2) manifold plus four large space/time dimensions. The study of such theories is ongoing. The compactified theories are testable in the traditional way of testing physics theories for four centuries. In fact, they're testable in the same sense as Newton's second law, $F = ma$. $F = ma$ is not testable in general but only for one force at a time—for a given force and mass object, you calculate the predicted acceleration and measure it. Similarly, the form the small extra dimensions take for compactified M/string theories leads to calculable and testable predictions.

A nice example of how the string theories may help comes from the Higgs boson mass. In the Standard Model, the Higgs boson mass cannot be predicted at all. The extension of the Standard Model to the theory called the Supersymmetric Standard Model predicts an upper limit on the Higgs boson mass but cannot make an accurate prediction of the mass. Compactified M-theory allows a prediction (made by me with students and colleagues) with an accuracy of a few percent, in 2011, before the CERN measurements and confirmed by subsequent data.

If we want to understand and explain our world, going beyond even a full mathematical description, we should take seriously, and work on, 10-D string theories or 11-D M-theory, compactifying them to our apparent 4-D world. People often say that string theories are complicated. Actually, compactified M/string theories seem to be the simplest theories that could encompass and integrate all the phenomena of the physical world into one coherent mathematical theory.

THE "NATURALNESS" ARGUMENT

PETER WOIT

Mathematical physicist, Columbia University; author, Not Even Wrong: The Failure of String Theory and the Search for Unity in Physical Law

For anyone currently thinking about fundamental physics, this latest *Edge* Question is easy, with an obvious answer: string theory. The idea of unifying physics by positing strings moving in ten space/time dimensions as fundamental entities was born in 1974 and became the dominant paradigm for unification from 1984 on. After forty years of research and tens of thousands of papers, what we've learned is that this is an empty idea. It predicts nothing about anything, since you can get pretty much any physics you want by appropriately choosing how to to make six of the ten dimensions invisible.

Despite this, proponents of the string-theory unification idea refuse to admit what has happened to it, often providing excellent examples of Max Planck's observation about what happens as scientists grow old while staying true to ideas that should have been discarded. Instead of retiring a failed idea, lately one hears instead that what needs to be retired are conventional ideas about scientific progress. According to string theorists, we live in an obscure corner of a multiverse where anything goes, and this "anything goes" fits right in with string theory, so fundamental physics has reached its end point.

The "string theory" answer to the 2014 *Edge* Question is, however, much too simplistic. String-theory unification has long been a moribund idea, but it's just part of a much larger circle of now failed ideas dating from the same period. These include so-called "grand unification" schemes proposing new forces and particles, generally invoking a new "supersymmetry" that relates known forces and particles to unseen "superpartners." Besides finding the predicted Higgs particle, the other great discovery of the Large Hadron Collider has been that the superpartners predicted by many theorists aren't there.

The period around 1974 brought us not only string theory, grand unification, and supersymmetry but also something called the "naturalness" argument. The idea here is that our best model of particle physics, the Standard Model, is just an "effective theory," an approximation, valid only at observable distance scales. The late Kenneth Wilson taught us how to use the "renormalization group" not only to extrapolate the behavior of a theory to short distances we can't observe but also how to run this backward, finding an effective theory for a fundamental theory defined at unobservably small distances. In a technical sense, "natural" theories are those where what we see is insensitive to the details of what happens at short distances. "Naturalness" became part of the speculative picture born in the 1974 era: Complicated new physics, involving unobserved strings and superpartners, could be postulated at very short distances, with a "natural" theory being all that's visible to us. In this picture, it's technical "naturalness" that ensures that we can't see any of the complexities introduced by unobservably small strings or superpartners.

Wilson was among the first to point out that the Standard Model is mostly "natural" but not entirely so, due to the behavior of the Higgs particle. At first he argued that this meant that at LHC energies we should see not the Higgs but something different. Fans of superpartners argued that such particles had to exist at roughly the same energy as the Higgs, since—if so—

they could be used to cancel the "unnaturalness." Long before the LHC turned on, Wilson had retracted this argument as a blunder, deciding that there was no good reason not to see an "unnatural" Higgs. The sensitivity of its behavior to what happens at very short distances is not a good argument against it, since we simply don't know what's going on at such distances.

The observation of the Higgs, but no superpartners, at the LHC has caused great consternation among theorists. Something has happened that shouldn't have been possible, according to the forty-year-old reasoning now embedded in textbooks. Arguments are being made that this is yet more evidence for the multiverse. In this "anthropic" view, anything goes at short distances for bubble universes elsewhere in the multiverse, but we see something "unnaturally" simple in *our* bubble universe, because otherwise we wouldn't be here. The rise of such reasoning shows that sending the naturalness argument into retirement (along with the epicyclic complexity of strings and superpartners) is long overdue.

THE COLLAPSE OF THE WAVE FUNCTION

FREEMAN DYSON

Professor emeritus of physics, Institute for Advanced Study; author, A Many-colored Glass: Reflections on the Place of Life in the Universe

Fourscore and eight years ago, Erwin Schrödinger invented wave functions as a way to describe the behavior of atoms and other small objects. According to the rules of quantum mechanics, the motions of objects are unpredictable. The wave function tells us only the probabilities of the possible motions. When an object is observed, the observer sees where it is, and the uncertainty of the motion disappears. Knowledge removes uncertainty. There is no mystery here.

Unfortunately, people writing about quantum mechanics often use the phrase "collapse of the wave function" to describe what happens when an object is observed. This phrase gives a misleading idea that the wave function itself is a physical object. A physical object can collapse when it bumps into an obstacle. But a wave function cannot be a physical object. A wave function is a description of a probability, and a probability is a statement of ignorance. Ignorance is not a physical object, and neither is a wave function. When new knowledge displaces ignorance, the wave function does not collapse; it merely becomes irrelevant.

QUANTUM JUMPS

DAVID DEUTSCH

Physicist, University of Oxford; author, The Beginning of Infinity; *recipient, Edge of Computation Science Prize*

The term "quantum jump" has entered everyday language as a metaphor for a large, discontinuous change. It has also become widespread in the vast but sadly repetitive landscape of pseudoscience and mysticism.

The term comes from physics and is indeed used by physicists (though rarely in published papers). It evokes the fact that mutually distinguishable states in quantum physical systems are always discrete. Yet there's no such phenomenon in quantum physics as a "quantum jump." Under the laws of quantum theory, change is always continuous in both space and time. OK, maybe some physicists still subscribe to an exception to that: namely, the so-called "collapse of the wave function" when an object is observed by a conscious observer. But that nonsense is not the nonsense I'm referring to here. I'm referring to misconceptions even about the submicroscopic world—like "When an electron in a higher-energy state undergoes a transition to a lower energy level, emitting a photon, it quantum-jumps from one discrete orbit to another without passing through intermediate states."

Even worse: "When an electron in a tunnel diode approaches the barrier that it does not have enough energy to penetrate (so that under classical physics it would bounce off), the quantum phenomenon of tunneling allows it to appear mysteriously on the other side without ever having been in the region where it would have negative kinetic energy."

The truth is that the electron in such situations does not have a single energy or position but a range of energies and positions, and the allowed range itself can change with time. If the whole range of energies of a tunneling particle were below that required to surmount the barrier, it would indeed bounce off. And if an electron in an atom really were at a discrete energy level, and nothing intervened to change that, then it would never make a transition to any other energy.

Quantum jumps are an instance of what used to be called "action at a distance"—something at one location having an effect, not mediated by anything physical, at another location. Newton called this idea "so great an Absurdity that I believe no Man who has in philosophical Matters a competent Faculty of thinking can ever fall into it."[c] And the error has analogs in fields quite distant from classical and quantum physics. For example, in political philosophy the "quantum jump" is called "revolution," and the absurd error is that progress can be made by violently sweeping away existing political institutions and starting from scratch. In the philosophy of science, it is Thomas Kuhn's idea that science proceeds via revolutions—that is, victories of one faction over another, both of which cannot alter their respective "paradigms" rationally. In biology, the "quantum jump" is called saltation, the appearance of a new adaptation from one generation to the next, and the absurd error is called saltationism.

Newton was wrong that there's a maximum size of error that competent people can fall into, but right that this particular one is severe. All those versions of it are mistaken for a single reason: They all require information of the requisite kind to appear from nowhere. In reality, the space on the far side of the barrier cannot "know" that an electron, and not a proton or a bison, must appear there, until some physical change, originating at the electron, reaches it. The same holds when it isn't a spatial gap but a directly informational one: Political institutions and biological adaptations instantiate information—knowledge—about how a complex system can better meet the challenges facing it, and knowledge can be created only by processes of piecemeal variation and selection. And Kuhn's vision cannot explain how science has in fact been delivering knowledge about physical reality at an ever accelerating rate.

Quantum jumps in all these fields represent a retreat from explanation and therefore, in effect, an appeal to the supernatural. They all have the logic of the Sidney Harris cartoon "Then a Miracle Occurs," depicting a mathematician with a gap in his proof. As Richard Dawkins puts it, "Saltationism is creationism." And in all cases the reality that fills the gap, the idea that truly explains the phenomenon, is much more interesting and delightful than any faith in its mystery could be.

CAUSE AND EFFECT

W. DANIEL HILLIS

Physicist, computer scientist, co-chairman of Applied Minds, LLC; author, The Pattern on the Stone

We humans are fundamentally storytellers. We like to organize events into chains of causes and effects that explain the consequences of our actions. We like to assign credit and blame. This makes sense from an evolutionary standpoint. The ultimate job of our nervous system is to make actionable decisions, and predicting the consequences of those decisions is important to our survival.

Science is a rich source of powerful explanatory stories. For example, Newton explained how a force causes a mass to accelerate. This gives us a story of how an apple drops from a tree and a planet circles the sun. It allows us to decide how hard the rocket engine needs to push to get it to the moon. Models of causation let us design complex machines, like factories and computers, that have fabulously long chains of causes and effects. They convert inputs into the outputs we want.

It's tempting to believe that our stories of causes and effects are how the world works. Actually, they're just a framework we use to manipulate the world and construct explanations for the convenience of our own understanding. For example, Newton's equation $F = ma$ doesn't really say that force causes acceleration, any more than it says that mass causes force. We tend to think of force as contingent, because we often have the choice as to whether to apply it or not. On the other hand, we tend to think of mass as not being under our control. Thus, we personify nature, imagining it almost as if natural forces were deciding to push on masses. It's much harder for us to imagine accelerations deciding to cause mass, so we tell the story a certain way. We credit gravitational force for keeping the planets orbiting the sun and blame it for pulling the apple down from the tree.

This convenient personification of nature helps us use our mental storytelling machinery to explain the natural world. The cause-and-effect paradigm works particularly well when science is used for engineering, to arrange the world for our convenience. In this case, we can often set things up so that the illusion of cause-and-effect is almost a reality. The computer is a perfect example. The key to what makes a computer work is that the inputs affect the outputs but not vice versa. The components used to construct the computer are constructed to create that same one-way relationship. These components, such as logic gates, are specifically designed to convert contingent inputs into predictable outputs. In other words, the logic gates of the computer are constructed to be atomic building blocks of cause and effect.

The notion of cause and effect breaks down when the parts we would like to think of as outputs affect the parts we would prefer to think of as inputs. The paradoxes of quantum mechanics are a perfect example of this, where our mere observation of a particle can "cause" a distant particle to be in a different state. Of course there's no real paradox here, there's just a problem with trying to apply our storytelling framework to a situation where it doesn't match.

Unfortunately, the cause-and-effect paradigm doesn't fail just at the quantum scale. It also falls apart when we try to use causation to explain complex dynamical systems like the biochemical pathways of a living organism, the transactions of an economy, or the operations of the human mind. These systems all have patterns of information flow that defy our tools of storytelling. A gene does not "cause" a trait like height or a disease like cancer. The stock market did not go up "because" the bond market went down. These are just our feeble attempts to force a storytelling framework onto systems that don't work like stories. For such complex systems, science will need more powerful explanatory tools, and we'll learn to accept the limits of our old methods of storytelling. We'll come to recognize that causes and effects don't exist in nature—that they're just convenient creations of our minds.

RACE

NINA JABLONSKI
Biological anthropologist, paleobiologist; Distinguished Professor of Anthropology, Pennsylvania State University

Race has always been a vague and slippery concept. In the mid-18th century, European naturalists such as Linnaeus, Comte de Buffon, and Johannes Blumenbach described geographic groupings of humans who differed in appearance. The philosophers David Hume and Immanuel Kant both were fascinated by human physical diversity; they held that extremes of heat, cold, or sunlight extinguished human potential. Hume, writing in 1748, contended that "there was never a civilized nation of any complexion other than white."

Kant felt similarly. He was preoccupied with questions of human diversity throughout his career and wrote at length on the subject in a series of essays beginning in 1775. He was the first to name and define the geographic groupings of humans as "races" (in German, *Rassen*). There were four of them—characterized by skin color, hair form, cranial shape, and other anatomical features, and also by their capacity for morality, self-improvement, and civilization. And they were arranged hierarchically, with only the European race, in his estimation, being capable of self-improvement.

Why did the scientific racism of Hume and Kant prevail in the face of the logical and thoughtful opposition of Johann Gottfried von Herder and others? Perhaps because Kant was recognized as a great philosopher in his own time, and his status rose in the 19th century as copies of his major philosophical works were widely distributed and read. Some of his supporters agreed with his racist views; some apologized for them; most commonly, many just ignored them. Moreover, racism—which diminished or denied the humanity of non-Europeans, especially Africans—bolstered the transatlantic slave trade, which had become the overriding engine of European economic growth. This view was augmented by biblical interpretations popular at the time which depicted Africans as destined for servitude.

Skin color, as the most noticeable racial characteristic, was associated with a nebulous assemblage of opinions and hearsay about the inherent natures of the various races. Skin color stood for morality, character, and the capacity for civilization; it became a meme. The 19th and early 20th centuries saw the rise of "race science." The biological reality of races was confirmed by new types of scientific evidence amassed by new types of scientists—notably anthropologists and geneticists. This era witnessed the birth of eugenics and its offspring, the concept of racial purity. The advent of Social Darwinism further reinforced the notion that the superiority of the white race was part of the natural order. The fact that all peoples are products of complex genetic mixtures resulting from migration and intermingling over thousands of years was not admitted by the racial scientists, nor by the scores of eugenicists who campaigned on both sides of the Atlantic for the improvement of racial quality.

The mid-20th century witnessed the continued proliferation of scientific treatises on race. By the 1960s, however, two factors were contributing to the demise of the concept of biological races. One was the increased rate of study of the physical and genetic diversity of human groups worldwide; the other was the emergence of the civil rights movement in the United States and elsewhere. Before long, influential scientists were denouncing studies of "race" because races themselves could not be scientifically defined. Where scientists looked for sharp boundaries between groups, none could be found. But despite these major shifts in scientific thinking, the sibling concepts of human races and a color-based hierarchy of races remained firmly established in mainstream culture. Racial stereotypes were potent and persistent, especially in the United States and South Africa, where subjugation and exploitation of dark-skinned labor had been the cornerstone of economic growth.

After its scientific demise, race remained as a name and concept but gradually came to stand for something quite different. Today many people identify themselves as belonging to one or another racial group regardless of what science may say about the nature of race. The shared experiences of members of such groups create powerful social bonds. For many people, including many scholars, the concept of race, while no longer biological, has become a mélange of social categories of class and ethnicity.

Clinicians continue to map observed patterns of health and disease onto old racial concepts such as "White," "Black" (or "African American"), "Asian," and so on. Even after it has been shown that many diseases (adult-onset diabetes, alcoholism, high blood pressure, to name a few) show apparent racial patterns because people share similar environmental conditions, groupings by race are maintained. The use of racial self-categorization in epidemiological studies is defended and even encouraged. Medical studies of health disparities between "races" become meaningless when sufficient variables—such as differences in class, ethnic social practices, and attitudes—are taken into account.

Race's latest makeover arises from genomics and mostly in biomedical contexts. The sanctified position of medical science in the popular consciousness gives the race concept renewed esteem. Racial realists marshal genomic evidence to support the hard biological reality of racial difference, while racial skeptics see no racial patterns. What's clear is that people are seeing what they want to see, constructing studies to provide the outcomes they expect. In *Race Decoded: The Genomic Fight for Social Justice* (2012), the University of California sociologist Catherine Bliss cogently characterizes race today as "a belief system that produces consistencies in perception and practice at a particular social and historical moment."

Race has a hold on history but no longer has a place in science. The sheer instability and potential for misinterpretation render race useless as a scientific concept. Inventing new vocabularies to deal with human diversity and inequity won't be easy, but it must be done.

ESSENTIALISM

RICHARD DAWKINS

Evolutionary biologist; Emeritus Professor of the Public Understanding of Science, Oxford; author, The Magic of Reality

Essentialism—what I've called "the tyranny of the discontinuous mind"—stems from Plato, with his characteristically Greek geometer's view of things. For Plato, a circle, or a right triangle, were ideal forms, definable mathematically but never realized in practice. A circle drawn in the sand was an imperfect approximation to the ideal Platonic circle hanging in some abstract space. That works for geometric shapes like circles, but essentialism has been applied to living things, and Ernst Mayr blamed this for humanity's late discovery of evolution—as late as the 19th century. If, like Aristotle, you treat all flesh-and-blood rabbits as imperfect approximations to an ideal Platonic rabbit, it won't occur to you that rabbits might have evolved from a non-rabbit ancestor and might evolve into a non-rabbit descendant. If you think, following the dictionary definition of essentialism, that the *essence* of rabbitness is "prior to" the *existence* of rabbits (whatever "prior to" might mean, and that's a nonsense in itself), evolution is not an idea that will spring readily to your mind, and you may resist when somebody else suggests it.

Paleontologists will argue passionately about whether a particular fossil is, say, *Australopithecus* or *Homo*. But any evolutionist knows there must have existed individuals who were exactly intermediate. It's essentialist folly to insist on the necessity of shoehorning your fossil into one genus or the other.

There never was an *Australopithecus* mother who gave birth to a *Homo* child, for every child ever born belonged to the same species as its mother. The whole system of labeling species with discontinuous names is geared to a time slice, the present, in which ancestors have been conveniently expunged from our awareness (and "ring species" tactfully ignored). If by some miracle every ancestor were preserved as a fossil, discontinuous naming would be impossible. Creationists are misguidedly fond of citing "gaps" as embarrassing for evolutionists, but gaps are a fortuitous boon for taxonomists who, with good reason, want to give species discrete names. Quarreling about whether a fossil is "really" *Australopithecus* or *Homo* is like quarreling over whether George should be called "tall." He's five-foot-ten—doesn't that tell you what you need to know?

Essentialism rears its ugly head in racial terminology. The majority of "African Americans" are of mixed race. Yet so entrenched is the essentialist mindset that official U.S. forms require everyone to tick one race/ethnicity box or another—no room for intermediates. A different but also pernicious point is that a person will be called "African American" even if, say, only one of his eight great grandparents was of African descent. As Lionel Tiger put it to me, we have here a reprehensible "contamination metaphor." But I mainly want to call attention to our society's essentialist determination to dragoon a person into one discrete category or another. We seem ill equipped to deal mentally with a continuous spectrum of intermediates. We are still infected with the plague of Plato's essentialism.

Moral controversies such as those over abortion and euthanasia are riddled with the same infection. At what point is a brain-dead accident victim defined as "dead"? At what moment during development does an embryo become a "person"? Only a mind infected with essentialism would ask such questions. An embryo develops gradually from single-celled zygote to newborn baby and there's no one instant when "personhood" should be deemed to have arrived. The world is divided into those who get this truth and those who wail, "But there has to be *some* moment when the fetus becomes human." No, there really doesn't, any more than there has to be a day when a middle-aged person becomes old. It would be better—though still not ideal—to say the embryo goes through stages of being a quarter human, half human, three-quarters human. . . . The essentialist mind will recoil from such language and accuse me of all manner of horrors for denying the *essence* of humanness.

Evolution, too, like embryonic development, is gradual. Every one of our ancestors, back to the common root we share with chimpanzees and beyond, belonged to the same species as its own parents and its own children. And likewise for the ancestors of a chimpanzee, back to the same shared progenitor. We are linked to modern chimpanzees by a V-shaped chain of individuals who once lived and breathed and reproduced, each link in the chain being a member of the same species as its neighbors in the chain, no matter that taxonomists insist on dividing them at convenient points and thrusting discontinuous labels upon them. If all the intermediates, down both forks of the V from the shared ancestor, had happened to survive, moralists would have to abandon their essentialist, "speciesist" habit of placing *Homo sapiens* on a sacred plinth, infinitely separate from all other species. Abortion would no more be "murder" than killing a chimpanzee—or, by extension, any animal. Indeed an early-stage human embryo, with no nervous system and presumably lacking pain and fear, might defensibly be afforded less moral protection than an adult pig, which is clearly well equipped to suffer. Our essentialist urge toward rigid definitions of "human" (in debates over abortion and animal rights) and "alive" (in debates over euthanasia and end-of-life decisions) makes no sense in the light of evolution and other gradualistic phenomena.

We define a poverty "line": You are either "above" or "below" it. But poverty is a continuum. Why not say, in dollar-equivalents, how poor you actually are? The preposterous Electoral College system in U.S. presidential elections is another, and especially grievous, manifestation of essentialist thinking. Florida must go either wholly Republican or wholly Democrat—all twenty-nine Electoral College votes—even though the popular vote is a dead heat. But states should not be seen as *essentially* red or blue: They're mixtures in various proportions.

You can surely think of many other examples of "the dead hand of Plato," essentialism. It's scientifically confused and morally pernicious. It needs to be retired.

HUMAN NATURE

PETER RICHERSON
Distinguished Professor emeritus, Department of Environmental Science and Policy, University of California, Davis; author, Not by Genes Alone

The concept of human nature has considerable currency among evolutionists who are interested in humans. Yet when examined closely, it's vacuous. Worse, it confuses the thought processes of those who attempt to use it. Useful concepts are those that cut nature at its joints. Human nature smashes bones.

Human nature implies that our species is characterized by a common core of features that define us. Evolutionary biology teaches us that this sort of essentialist concept of species is wrong. A species is an assemblage of variable individuals, albeit individuals sufficiently genetically similar that they can successful interbreed. Most species share most of their genes with ancestral and related species, as we do with other apes. In most species, ample genetic variation ensures that no two individuals are genetically identical. Many species contain geographically structured genetic variation, as modern humans do. A few tens of thousands of years ago, our genus seems to have comprised a couple of African "species" and three Eurasian ones, all of which interbred enough to leave traces in living genomes. Most species, and the populations of which they're composed, are relentlessly evolving. The human populations that have adopted agriculture in the Holocene have undergone a wave of genetic changes to adapt to a diet rich in starchy staples and other agricultural products and an environment rich in epidemic pathogens taking advantage of dense, settled human populations. Some human populations today are subject to new selective pressures, owing to "diseases of abundance." The evolution of resistance to such diseases is detectable. Some geneticists argue that genes affecting our behavior have come under recent selection to adapt to life in complex societies.

The concept of human nature causes people to look for explanations under the wrong rock. Take the most famous human-nature argument: Are people by nature good or evil? In recent years, experimentalists have conducted tragedy-of-the-commons games and observed how people solve the tragedy (if they do). A typical finding is that roughly a third of the participants act as selfless leaders, using whatever tools the experimenters make available to solve the dilemma of cooperation, roughly a tenth are selfish exploiters of any cooperation that arises, and the balance are guarded cooperators with flexible morals. This result comports with everyone's personal experience: Some people are routinely honest and generous, a few are downright psychopathic, and many people fall somewhere in between. Human society would be entirely different if this weren't so. The human-nature debate on the topic is sterile, because it ignores something we all know, if we stop to think about it.

Darwin's great contribution to biology was to abandon essentialism and focus on variation and its transmission. He made remarkable progress, even though organic inheritance was a black box in his day. He also got the main problem of human variability right. In *The Descent of Man*, he argued that humans were biologically a rather ordinary species with a rather ordinary amount of geographical variation. Yet in many ways the amount of human behavioral variation is far outside the range of that of other species. The Fuegans who adapted to a hunter-gatherer life on the Strait of Magellan were sharply different from a leisured gentleman naturalist from Shrewsbury, but these differences owe mainly to different customs and traditions, not organic differences. Darwin also realized that the evolution of traditions responded to selective processes other than natural selection. Traditions are shaped by human choices, a little like the artificial selection of domesticates, with natural selection playing a subordinate role.

In his *Biographical Sketch of an Infant,* Darwin described how readily children learn from their caregivers. The inheritance of traditions, customs, and language is relatively easy to observe with the tools of a 19th-century naturalist, compared to the intricacies of genetic inheritance, which is still yielding fundamental secrets to the high-tech tools of molecular biology. Recent work on the mechanisms underlying imitation and teaching has begun to reveal the more deeply hidden cognitive components of these processes, and the results underpin Darwin's phenomenological account of tradition acquisition and evolution.

In no field is the deficiency of the human-nature concept better illustrated than in its use in trying to understand learning, culture, and cultural evolution. Human-nature thinking leads to the conclusion that causes of behavior can be divided into nature and nurture. Nature is conceived of as causally prior to nurture, both in evolutionary and developmental time. What evolves is nature, and cultural variation, whatever it is, has to be the causal handmaiden of nature. This is simply counterfactual. If the dim window that stone tools give us doesn't lie, culture and cultural variation have been fundamental adaptations of our lineage perhaps going back to late australopiths. The elaboration of technology over the last 2 million years has roughly paralleled the evolution of larger brains and other anatomical changes. We have clear examples of cultural changes driving genetic evolution, such as the evolution of dairying driving the evolution of adult lactase persistence. Socially learned technology could have been doing similar things all throughout the last 2 million years. The human capacity for social learning develops so early in the first year of life that developmentalists have had to design very clever experiments to probe what infants are learning months before language and precise imitative behavior exist. At least from twelve months onward, social learning begins to transmit the discoveries of cultures to children with every opportunity for these discoveries to interact with gene expression. In autistic children, this social-learning mechanism is more or less severely compromised, leading to more or less severely "developmentally disabled" adults.

Human culture is best conceived of as a part of human biology, like our bipedal locomotion. It's a source of variation we've used to adapt to most of the world's terrestrial and amphibious habitats. The human-nature concept, like essentialism more generally, makes it impossible to think straight about human evolution.

THE *URVOGEL*

JULIA CLARKE

Associate professor, John A. Wilson Centennial Fellow in Vertebrate Paleontology, Jackson School of Geosciences, University of Texas, Austin

I would like to put to rest the notion that evolution as a process should conform to words and concepts we find familiar, comfortable, and perhaps even universal. More immediately, I'd like to stop having always to explain whether or not each new feathered dinosaur specimen we discover was a bird.

In many ways it's an understandable question. Most scientists have accepted for years that living birds are one lineage of dinosaurs. The idea that dinosaurs live on in birds has even crept into popular consciousness, via *Jurassic Park*. So perhaps it's not surprising that when scientists discover a new feathered dinosaur, people—including scientists and science journalists—often want to know, "Did it fly?" Consider the so-called *Urvogel*—the first-discovered feathered dinosaur, *Archaeopteryx*. Debate continues in the scientific literature: Was it a bird?

As a paleontologist working on the evolution of living birds, I find myself having this exchange over and over again. For example, I describe a small feathery species newly uncovered from the fossil record. After detailing its known features, I might note that it may have had some form of aerial locomotion. There is inevitably a pause. Then comes the question: "OK, but was it a bird?" Impatient with scientists and their endless modifiers and complex phrasing, the asker wants to get this story clear. ("OK, but did it fly? Tell it to me straight.")

The questions may sound innocent enough and perhaps are natural to ask. However, although they seem like scientific questions, they mostly aren't. They concern primarily what we want to identify as part of a class of entities (birds) and what part of that class (flighted). We might think we have these distinctions straight today, but try looking back through a dirty lens at life more than 100 million years ago.

Paleontologists use the shape and form of bone as well as, in rare cases, feathery impressions to track the ecologies of the long dead. To do this, they use data on form/function relationships in the living. This task is itself difficult and ongoing. But what's more difficult is to translate combinations of structures not present in any living species into an understanding of how the animal under study moved. For example, flighted living birds have a joint between the scapula and coracoid where the upper arm bone, humerus, meets the pectoral girdle. Yet we have species in the fossil record with feathered forelimbs of impressive span (shall we call them wings?) but lacking this kind of joint articulation. Subtle features of the feathers and their relative proportions may differ from any living bird. Is this creature a bird?

How did it move? Did it have a kind of sustained flapping flight but unlike that in any extant species? If we could time-travel back to a Cretaceous forest, would we call this movement "flight"? What if a species beat its wings only briefly to move from branch to branch? What if it utilized these "wing" beats to climb trees or jump? What if it was only volant as a juvenile, while as a large-bodied adult it maintained feathery forelimbs for signaling to a mate but flew no longer?

All these hypotheses have been put forward, and all may have been true for different denizens of Jurassic and Cretaceous environments. We can debate whether these creatures flew and whether or not they were birds by contemporary definition, but in doing so we risk losing sight of the bigger scientific questions. All too quickly, we can fall into a rabbit hole of defining (and defending) terms, when we'd do better to seek a more precise understanding of the emergence, the relative evolutionary first appearances, of the many features comprising the flight apparatus in living birds.

Feathers make their first appearance in taxa that could not have been volant as adults. Precursors to feathers, simple filaments, are found in tyrannosaurids and an array of other relatives of living birds. While hundreds of characteristics of bone and feather have revealed these deep genealogical relationships within dinosaurs, we still seem to be searching to pin "bird" and "flight" to single characters.

I'm not the first to remark that the debate over what to call a "bird" and what to term "flight" is not useful and is actually at odds with evolutionary thinking. But I'm surprised by the persistence of this debate even among specialists. For example, exchanges over how to apply the formal taxonomic name *Aves* are ongoing. While events unfolding in deep time via evolutionary processes are arguably the least likely candidates for dichotomous or categorical thinking, this mode of thought runs rampant and engenders false controversies that obscure interesting questions. Tracking the more complex pattern of asynchronous change in many novel traits is what will inform generalities about how the evolution of shape and form may work.

The hypotheses we investigate should be arrayed relative to one another in relationships other than opposition; however, often the categories we're comfortable with will artificially organize them into this apparent relationship. Indeed, across science we have many *Urvogels*—lingering evidence of similarly strong collective cognitive investment in the existence of classes of entities we consider intuitive and natural. These can hold us back.

NUMBERING NATURE

KURT GRAY

Assistant professor of social psychology, University of North Carolina, Chapel Hill

> *"I was much struck how entirely vague and arbitrary is the distinction between species and varieties."* —Darwin, *On the Origin of Species* (1859)

For centuries, there was one way to bring order to the vastness of biological diversity—Linnaean classification. In the 18th century, Carl Linnaeus devised a method for dividing species based on their description: Do they look the same? Do they behave the same? With Linnaean classification, you could divide the natural world into discrete kinds. You could count them, saying confidently, "There are two species of elephants" or "There are four kinds of bears." Some psychologists seek to bring the same order to the mind, claiming that "There are six emotions," "There are five types of personality," or "There are three moral concerns." These psychologists are inspired by the precision, order, and neatness of Linnaeus' ideas—the only problem is that Linnaeus was wrong.

Linnaeus lived about 100 years before Darwin introduced his theory of evolution by natural selection, and he long believed that species were fixed and unchangeable. His religious roots led him to see species as a product of Divine Providence, and his job was simply to catalog these distinct kinds, taking as his motto "God created, Linnaeus organized." If God created a certain number of distinct species, cataloging and counting them made sense. It was meaningful to ask "How many kinds of salamanders did God create?"

Evolution, however, destroyed the sanctity of species. Species were not created whole from the Beginning but instead emerged over time, through the repetition of a simple algorithm: heredity, mutation, and selection. Evolution showed that a dizzying diversity of life—from viruses to cacti to humans—were explained through a basic set of common processes expressed in different environments. This common process means that lines between species are more in the minds of humans than in nature, with many intermediate animals (lungfish, for example) and hybrids (ligers, for example) that defy easy categorization. Moreover, in geological time these divisions are even more arbitrary, with species diverging and converging as continents separate and collide.

Biology has realized that species are not reflections of eternal Divine order but simply a useful way to intuitively organize the world. Unfortunately, psychology lags behind. Many psychologists believe that the mental world is fixed and countable, that the appearance of mental states reflects a deeper essence. Introductory psychology textbooks contain numbered lists of psychological species—five kinds of human needs, six basic emotions, three moral concerns, three kinds of love, three parts of the mind—with these lists depending primarily upon the intuitions of whoever is doing the counting.

Like those of Linnaeus in the 18th century, these intuitive taxonomies were once the best we could do, because psychology lacked an understanding of basic psychological processes. However, social cognition and neuroscience has revealed these processes and found that diverse mental experiences—emotion, morality, motivation—are combinations of more basic affective and cognitive processes. This research suggests that psychological states are not firmly demarcated "things" with enduring essences but are instead fuzzy constructs emerging from common psychological processes expressed across different environments.

Just as evolution can create a multitude of species by expressing a common process in specific environments, so too can the mind create a multitude of mental species. You can no more count emotions or moral concerns than snowflakes or colors. To be sure, there are descriptive similarities and differences across instances, but groupings are arbitrary and rest heavily on the intuition of researchers. This is why scientists can never agree on the fundamental number of anything; one scientist may divide a mental experience into three, another into four, and another into five.

It's time for psychology to abandon the enterprise of numbering nature and recognize that psychological species are neither distinct nor real. Biology has long recognized the arbitrary and constructed nature of species; why are we more than 200 years behind? The likely answer is that people—even including psychologists and philosophers—believe that intuitions, as products of the mind, are accurate reflections of its structure. Unfortunately, decades of research demonstrate the flaws of intuitive realism, revealing that intuitions about the mind are poor guides to underlying psychological processes.

Psychologists must move from counting to combining. *Counting* is simply describing the world—one psychologist's intuitive ordering of mental states in one culture, at one time. *Combining* seeks to find basic psychological elements and discover how they interact to create the mental world. In biology, counting asks, "How many kinds of salamanders are there?" whereas combining asks, "What processes lead to salamander diversity?" Counting is bound to a specific environment and time, whereas combining recognizes these factors as processes themselves. Psychology must follow biology and move from numbering individual species to exploring underlying systems.

This process has already begun. Thomas Insel, the head of the National Institute of Mental Health, has prioritized systems over species in psychopathology research. He rejects the utility of the *Diagnostic and Statistical Manual of Mental Disorders,* suggesting that intuitive taxonomies obscure underlying processes of psychopathology and impede the discovery of treatments. NIMH funds proposals that examine the underlying affective, conceptual, and neurological systems, which may explain why the "distinct" disorders of depression and anxiety are so often comorbid, and why selective serotonin reuptake inhibitors (SSRIs) appear to help diverse disorders. Psychopathology doesn't easily fit into categories, and neither do other psychological phenomena.

Of course, we shouldn't throw out the baby with the bathwater. It's still necessary to catalog the natural world to allow meaningful discussion. Even in biology, where the power of the process of evolution is undisputed, most acknowledge the utility of Linnaeus' system and continue to use the names he provided centuries ago. But the key is not to confuse human constructions with natural order. What's useful to humans isn't necessarily true of nature. Intuitive taxonomies are a necessary first step in psychological science, but even Linnaeus, as he learned more about the world, recognized the arbitrariness of his system and the species he labeled. It's time for psychology to recognize this fact as well and leave Linnaeus and the 18th century behind.

HARDWIRED = PERMANENT

MICHAEL SHERMER

Founding publisher, Skeptic *magazine; monthly columnist,* Scientific American*; author,* The Believing Brain

The scientific idea that a trait or characteristic of an organism that's "hardwired" means it's a permanent feature should be retired. Case in point: God and religion.

Ever since Charles Darwin theorized, in his 1871 book *The Descent of Man,* that "a belief in all-pervading spiritual agencies seems to be universal" and therefore an evolved characteristic of our species hardwired into our brains, scientists have been running experiments and conducting surveys to show why God won't go away. Anthropologists have found such human universals as specific supernatural beliefs about death and the afterlife, fortune and misfortune, and especially magic, myths, rituals, divination, and folklore. Behavioral geneticists report from twin studies—most notably, of twins separated at birth and raised in different environments—that between 40 and 50 percent of the variance of God beliefs and religiosity are genetic. Some scientists have even claimed to have found a "God gene" (or more precisely, a "God gene complex") that leads humans to have a need for spiritual transcendence and belief in a higher power of some kind. Even specific elements of religious stories—such as a destructive flood, a virgin birth, miracles, a resurrection from the dead—seem to appear independently, over and over again throughout history, in a wide variety of cultures, implying that there's a hardwired component to religion and God beliefs. I have held this theory myself. Until now.

If and when we establish a permanent colony on Mars, if its members consist of nonbelieving scientists with a purely secular worldview, it would be interesting to check in ten (or a hundred) generations to see whether God has returned. Until that experiment is conducted, however, we have to consider the results of natural experiments run here on Earth. In the Western world, for example, a 2013 survey of 14,000 people in 13 nations (Germany, France, Sweden, Spain, Switzerland, Turkey, Israel, Canada, Brazil, India, South Korea, and the U.K. and U.S.), conducted by the German pollster Bertelsmann Stiftung for their Religion Monitor, found that most of those countries showed a declining trend in religiosity and belief in God, especially among the youth. In Spain, for example, 85 percent of respondents over the age of forty-five report being moderately to very religious, but only 58 percent of those under twenty-nine so report. In Europe in general, only 30 to 50 percent say that religion is important in their lives, and in many European countries less than a third say they believe in God.

Even in the *über*-religious United States, the pollsters found that 31 percent of Americans say they're "not religious or not very religious." This finding confirms those of a 2012 Pew Forum survey: that the fastest-growing religious cohort in the U.S. are the "None"s (those with no religious affiliation) at 20 percent (33 percent of adults under thirty), broken down into atheists and agnostics at 6 percent and the unaffiliated at 14 percent. The raw numbers are stunning: With the U.S. adult population (age eighteen and over) at 240 million, this translates into 48 million "None"s, or 14.4 million atheists/agnostics and 33.6 million unaffiliated. There were also generational differences revealing a significant trend toward unbelief, with the "Greatest" generation (born 1913–1927) at 5 percent, the "Silent" generation (born 1928–1945) at 9 percent, the "Boomers" (born 1946–1964) at 15 percent, the "GenXers" (born 1965–1980) at 21 percent, the "Older Millennials" (born 1981–1989) at 30 percent, and the "Younger Millennials" (born 1990–1994) at 34 percent.

At this rate, I project that the "None"s will reach 100 percent in the year 2220.

It's time for scientists to retire the theory that God and religion are hardwired into our brains. Like everyone else, scientists are subject to cognitive biases that tilt their thinking toward trying to explain common beliefs, so it's good for us to take the long view and compare today with, say, half a millennium ago, when God beliefs were virtually 100 percent, or with the hunter-gatherer tribes of our Paleolithic ancestors who, while employing any number of superstitious rituals, didn't believe in a God or practice a religion that even remotely resembles the deities or religions of modern peoples.

This indicates that religious faith and belief in God is a by-product of other cognitive processes (for example, agency detection) and cultural propensities (the need to affiliate) that, while hardwired, can be expunged through reason and science in the same manner as any number of other superstitious rituals and supernatural beliefs once held by the most learned scholars and scientists of Europe five centuries ago. For example, at that time the prevailing theory to explain crop failures, weather anomalies, diseases, and various other maladies and misfortunes was witchcraft, and the solution was to strap women to pyres and torch them to death. Today, no one in his right mind believes this. With the advent of a scientific understanding of agriculture, climate, disease, and other causal factors—

including the role of chance—the witch theory of causality fell into disuse.

So it has been and will continue to be with other forms of the hardwired = permanent idea, such as violence. We may be hardwired for violence, but we can attenuate it considerably through scientifically tested methods. Thus, for my test case here, I predict that in another 500 years the God-theory of causality will have fallen into disuse and the 21st-century scientific theory that God is hardwired into our brains as a permanent feature of our species will be retired.

THE ATHEISM PREREQUISITE

DOUGLAS RUSHKOFF

Media analyst; documentary writer; author, Present Shock

We don't need to credit an all-seeing God with the creation of life and matter to suspect that something wonderfully strange is going on in the dimension we call reality. Most of us living in it feel invested with a sense of purpose. Whether this directionality is a genuine, preexisting condition of the universe, an illusion perpetrated by DNA, or something that will one day emerge from social interaction has yet to be determined. At the very least, this means our experience and expectations of life can no longer be dismissed as impediments to proper observation and analysis.

But science's unearned commitment to materialism has led us into convoluted assumptions about the origins of spacetime, in which time itself simply must be accepted as a by-product of the Big Bang, and consciousness (if it even exists) as a by-product of matter. Such narratives follow information on its continuing evolution toward complexity, the Singularity, and robot consciousness—a saga no less apocalyptic than the most literal interpretations of biblical prophecy.

It's entirely more rational—and less steeped in storybook logic—to work with the possibility that time predates matter and that consciousness is less the consequence of a physical cause-and-effect reality than a precursor.

By starting with Godlessness as a foundational principle of scientific reasoning, we make ourselves unnecessarily resistant to the novelty of human consciousness, its potential continuity over time, and the possibility that it has purpose.

EVOLUTION IS "TRUE"

ROGER HIGHFIELD
Director, external affairs, Science Museum Group; coauthor (with Martin Nowak), SuperCooperators

Politicians, poets, philosophers, and the religious often like to talk about the truth. In contrast, most scientists would think it overblown to describe a field of research as being "true," though they do all seek the truth of mathematics. For example, quantum theory is true in the sense that experiment after experiment supports its predictions about how the world works, no matter how odd, unsettling, or counterintuitive it is. Similarly, when I studied chemistry at university I was never told about the "truth" of the Periodic Table, though I did marvel at how Mendeleev had glimpsed the electronic structure of atoms. But why do some biologists talk about the truth so much when it comes to evolution? After all, one can hardly say that everything written about evolution is "true." It's a mistake to counter irrational beliefs with rhetoric about the Truth.

Intelligent design and other creationist critiques have been easily shrugged off and the facts of evolution well established in the laboratory, fossil record, DNA record, and computer simulations. If evolutionary biologists are really Seekers of the Truth, they need to focus more on finding the mathematical regularities of biology, following in the giant footsteps of Sewall Wright, J. B. S. Haldane, Ronald Fisher, and so on.

The messiness of biology has made it relatively hard to discern the mathematical fundamentals of evolution. Perhaps the laws of biology are deductive consequences of the laws of physics and chemistry. Perhaps natural selection is not a statistical consequence of physics but a new and fundamental physical law. Whatever the case, those universal truths—"laws"—that physicists and chemists all rely upon appear relatively absent from biology.

Little seems to have changed from a decade ago, when the late and great John Maynard Smith wrote a chapter on evolutionary game theory for a book on the most powerful equations of science: His contribution included not a single equation. Yet there are already many mathematical formulations of biological processes, and evolutionary biology will truly have arrived the day that high school students learn the equations of life in addition to Newton's laws of motion.

Moreover, if physics is an example of what a mature scientific discipline should look like—one that doesn't waste time and energy combating the agenda of science-rejecting creationists—we also need to abandon the blind adherence to the idea that the mechanisms of evolution are truths that lie beyond discussion. Gravity, like evolution, exists, but Newton's view of gravitation was absorbed into another view that Einstein devised a century ago. Even today, however, there's debate about whether our understanding of gravity will have to be modified again, when we're finally enlightened about the nature of the dark universe.

THERE IS NO REALITY IN THE QUANTUM WORLD

ANTON ZEILINGER

Physicist, University of Vienna and Institute for Quantum Optics and Quantum Information; president, Austrian Academy of Sciences; author, Dance of the Photons

What ought to be abandoned is the idea that there's no reality in the quantum world. This idea probably arose for two reasons: (1) because we cannot always ascribe a precise value to a physical property, and (2) because within the wide spectrum of interpretations of quantum mechanics, some suggest that the quantum state does not describe an external reality but rather that the properties come about only in the mind of the observer, and therefore that consciousness plays a crucial role.

Let's consider the famous double-slit experiment. Such experiments or their equivalents have been performed not just with single photons (or any other kind of particles, like neutrons, protons, or electrons) but even with a beam of very large particles such as the spherical C_{60} or the C_{70} carbon molecules, known as buckyballs. Under the right experimental conditions, you observe a distribution of the buckyballs on the screen behind the two slits—a distribution that has maxima and minima, the interference pattern due to the interference of the probability waves passing through both slits. But (following Einstein in his famous debate with Niels Bohr) we might ask: "Through which slit does a particular buckyball molecule pass?" Isn't it natural to assume that each molecule has to pass through either one slit or the other?

Quantum physics tells us that this isn't a meaningful question. We cannot assign a well-defined position to the particle unless we perform an experiment allowing us to find out where it is. Before we do this, the position of the buckyball—and therefore which slit it passes through—is a concept devoid of meaning.

Suppose we now measure the position of a particular particle in the buckyball beam. Then we get an answer. We know where it is: It's either near one slit or near the other slit. In that case, position is an element of reality and we can say that quantum physics describes this reality. What's interesting is that if we have precise knowledge of one feature (namely, the particle's position), another kind of knowledge (namely, the one encoded in the interference pattern) is no longer well defined.

Where could consciousness come in here? Quantum mechanics tells us that the particle, before any observation is made of it, is in a superposition of passing through both slits. If we now place two detectors, one behind each slit, then either one detector or the other will register the particle. But quantum mechanics tells us that the measurement apparatus becomes entangled with the position of the particle and thus itself lacks well-defined classical features—at least in principle. This entanglement persists until an observer registers the result. So if we adopt that reasoning, it's the consciousness of the observer which makes reality happen.

But we needn't go that far. It's enough to assume that quantum mechanics just describes the *probabilities* of possible measurement results. Then making an observation turns potentiality into actuality—and, in our case, the position of the buckyball becomes a quantity that one can talk reasonably about. But whether it has a well-defined position or not, the buckyball exists. It is *real* in the double-slit experiment, even though we cannot assign to its position a well-defined value.

SPACETIME

STEVE GIDDINGS

Theoretical physicist, University of California, Santa Barbara

Physics has always been regarded as playing out on an underlying stage of space and time. Special relativity joined these into spacetime, and general relativity taught us that this spacetime itself bends and ripples—but it has remained part of the foundations of physics. However, the need to give a quantum-mechanical description of reality challenges the very notion that space and time are fundamental.

We specifically face the problem of reconciling the principles of quantum mechanics with the physics of gravity. At first, physicists believed this meant that spacetime could violently fluctuate to the point of losing meaning—though only at extremely short distances. But attempts to reconcile quantum principles with gravitational phenomena indicate a more profound challenge to the foundational role of spacetime. This comes to the fore when we study both black holes and the evolution of the universe. Spacetime structure is also seemingly problematic at very long distances.

Quantum mechanics appears to be an inevitable aspect of physics and is remarkably resistant to modification. If quantum principles govern nature, it seems likely that spacetime arises from more fundamentally quantum structures—and its emergence is, perhaps, roughly analogous to the emergence of fluid behavior from the interactions of atoms.

The problem with fundamental spacetime is even more strongly hinted at from many developing perspectives. Notable among these hints is the physics of black holes, where it appears that evolution that respects quantum principles must violate the classical spacetime dictum that information doesn't propagate faster than the speed of light. Something seems very wrong with the standard spacetime picture. Evidence mounts when we consider the large-scale structure of the universe, given quantum principles and the presence of dark energy. Here ultimately spacetime undergoes strong quantum fluctuations at very long scales and seems to lose meaning. More hints have come from candidate mathematical approaches to fluctuating spacetime.

The apparent need to retire classical spacetime as a fundamental concept is profound—and a clear successor is not yet in sight. Various approaches to the underlying quantum framework exist. Some show promise, but none yet clearly resolves our decades-old conundrums in black holes and cosmology. The emergence of such a successor will likely be a key element in the next major revolution in physics.

THE UNIVERSE

AMANDA GEFTER

Science writer; consultant, New Scientist*; author,*
Trespassing on Einstein's Lawn

Physics has a time-honored tradition of laughing in the face of our most basic intuitions. Einstein's relativity forced us to retire our notions of absolute space and time, while quantum mechanics forced us to retire our notions of pretty much everything else. Still, one stubborn idea has stood steadfast through it all: the universe.

Sure, our picture of the universe has evolved over the years—its history dynamic, its origin inflating, its expansion accelerating. It has even been downgraded to just one in a multiverse of infinite universes forever divided by event horizons. But still we've clung to the belief that here, as residents in the Milky Way, we all live in a single spacetime, our shared corner of the cosmos—our universe.

In recent years, however, the concept of a single shared spacetime has sent physics spiraling into paradox. The first sign that something was amiss came from Stephen Hawking's landmark work in the 1970s showing that black holes radiate and evaporate, disappearing from the universe and purportedly taking some quantum information with them. Quantum mechanics, however, is predicated upon the principle that information can never be lost.

Here was the conundrum. Once information falls into a black hole, it can't climb back out without traveling faster than light and violating relativity. Therefore, the only way to save it is to show that it never fell into the black hole in the first place. From the point of view of an accelerated observer who remains outside the black hole, that's not hard to do. Thanks to relativistic effects, from his vantage point, the information stretches and slows as it approaches the black hole, then burns to scrambled ash in the heat of the Hawking radiation before it ever crosses the horizon. It's a different story, however, for the inertial, infalling observer, who plunges into the black hole, passing through the horizon without noticing any weird relativistic effects or Hawking radiation, courtesy of Einstein's equivalence principle. For him, information better fall into the black hole or relativity is in trouble. In other words, in order to uphold all the laws of physics, one copy of the bit of information has to remain outside the black hole while its clone falls inside. Oh, and one last thing—quantum mechanics forbids cloning.

Stanford physicist Leonard Susskind eventually solved the information paradox by insisting that we restrict our description of the world to either the region of spacetime outside the black hole's horizon or to the interior of the black hole. Either one is consistent—it's only when you talk about both that you violate the laws of physics. This "horizon complementarity," as it became known, tells us that the inside and outside of the black hole are not part and parcel of a single universe. They are *two* universes, but not in the same breath.

Horizon complementarity kept paradox at bay until last year, when the physics community was shaken by a new conundrum more harrowing still—the so-called firewall paradox. Here, our two observers find themselves with contradictory quantum descriptions of a single bit of information, but now the contradiction occurs while both observers are still outside the horizon, before the inertial observer falls in. That is, it occurs while they're still supposedly in the same universe.

Physicists are beginning to think that the best solution to the firewall paradox may be to adopt "strong complementarity"—that is, to restrict our descriptions not merely to spacetime regions separated by horizons but to the reference frames of individual observers, wherever they are. As if each observer had his or her own universe.

Ordinary horizon complementarity has already undermined the possibility of a multiverse. If you violate physics by describing two regions separated by a horizon, imagine what happens when you describe *infinite* regions separated by *infinite* horizons! Now strong complementarity is undermining the possibility of a single, shared universe. On a glance, you'd think it would create its own kind of multiverse, but it doesn't. Yes, there are multiple observers, and yes, any observer's universe is as good as any other's. But if you want to stay on the right side of the laws of physics, you can talk only about one at a time. Which means, really, that only one *exists* at a time. It's cosmic solipsism.

Sending the universe into early retirement is a radical move, so it better buy us something pretty in the way of scientific advancement. I think it does. For one, it might shed some light on the disconcerting low-quadrupole coincidence—the fact that the cosmic microwave background radiation shows no temperature fluctuations at scales larger than 60 degrees on the sky, capping the size of space at precisely the size of our observable universe—as if reality abruptly stopped at the edge of an observer's reference frame.

More important, it could offer us a better conceptual grasp of quantum mechanics. Quantum mechanics defies understanding because it allows things to hover in superpositions of mutually exclusive states, as when a photon goes through this slit *and* that slit, or when a cat is simultaneously dead *and* alive. It balks at our Boolean logic; it laughs at the law of the excluded middle. Worse, when we actually observe something, the superposition vanishes and a single reality miraculously unfurls.

In light of the universe's retirement, this all looks slightly less miraculous. After all, superpositions are really superpositions of reference frames. In any single reference frame, an animal's vitals are well defined. Cats are only alive *and* dead when you try to piece together multiple frames under the false assumption that they're all part of the same universe.

Finally, the universe's retirement might offer some guidance as physicists push forward with the program of quantum gravity. For instance, if each observer has his or her own universe, then each observer has his or her own Hilbert space, his or her own cosmic horizon, and his or her own version of holography, in which case what we need from a theory of quantum gravity is a set of consistency conditions that can relate what different observers can operationally measure.

Adjusting our intuitions and adapting to the strange truths uncovered by physics is never easy. But we may just have to come around to the notion that there's my universe and there's your universe—but there's no such thing as *the* universe.

THE HIGGS PARTICLE CLOSES A CHAPTER IN PARTICLE PHYSICS

HAIM HARARI

Physicist; chair, Davidson Institute of Science Education; former president, Weizmann Institute of Science; author, A View from the Eye of the Storm

The discovery of the Higgs particle (aka the God particle, aka the Goddamn particle, according to Leon Lederman) allegedly closes the chapter of establishing the Standard Model of particle physics—or at least so we read in the newspapers and in the announcements from Stockholm. The introduction of this idea, five decades ago, was indeed an important landmark in the development of the Standard Model. But in reality it doesn't answer any of the remaining open questions, which have plagued the model for more than thirty years.

Nature has taught us that everything (not really; what about the dark matter and dark energy?) is made of six types of quarks (why six?) and six types of leptons (why six and why the same number?). They're arranged in a clear pattern, which replicates itself (why?) three times (why three?) in a precise manner. These dozen types of particles have positive or negative electric charges of exactly 0,1, 2, or 3 units in multiples of one-third of the electron charge (why always only those charges, no others, and why are quark charges even related to lepton charges?). The particle masses can be described only by approximately twenty free parameters, unrelated to one another, appearing to be taken from the results of some bizarre cosmic lottery, ranging over almost ten orders of magnitude.

Yes, the Higgs concept gives us a tantalizing mechanism by which these particles obtain a mass and are not massless. But this is what creates the problem. Why these masses? Who selected these numbers and why? Can it be that all of physics—and, indeed, all of science—is based on creating all the matter in the universe from a dozen objects with totally random mass values, while no one has the faintest idea about their origin?

These mysterious mass values allegedly reflect the strength in which the Higgs particle "couples" to the quarks and leptons. But that's like saying that the weights of a dozen people reflect the fact that when they step on the scale, these numbers appear there. Not a very satisfying explanation. The true puzzle of the Standard Model is, as always in physics, "What next?" Something must lie beyond it, solving the puzzle of the dark matter, dark energy, particle masses, and their simple, distinct, and repetitive systematic pattern.

The Higgs particle contributes absolutely nothing to the solution of these puzzles, unless the final answer is that the Higgs particle is indeed God's particle and it is God's will that the particle masses are these and no others. Or perhaps it is not one God but a dozen gods with diverse numerical tastes. The good news is that we still have some exciting discoveries ahead of us, deciphering the basic structure of all of matter, beyond the temporary picture offered by the Standard Model. We certainly do not yet have a Theory of Everything—not even close.

AESTHETIC MOTIVATION

SARAH DEMERS

Assistant professor of physics, Yale University

The Standard Model of particle physics has aesthetic shortcomings that leave us with questions: Why so many free parameters? Why not an elegant, single fundamental force to account for all forces? Why three generations of quarks and leptons? Now that we have a mechanism for how fundamental particles acquire mass, why do they have those particular couplings to the Higgs field, covering such a huge range of masses? Why the even more extreme range of strengths of the fundamental forces? The potential danger with each of these questions is the answer "That's just the way it is."

Besides these aesthetic concerns, we have contradictions between prediction and observation in the explored universe. We have not found a source of energy to fuel the accelerating expansion; there's insufficient baryonic matter to explain astronomical observations. We live in a large pocket of it that should not have survived annihilation. In fact, we see matter dominance everywhere we look—and no sufficient source of matter-versus-antimatter asymmetry to account for this. We may never access solutions to this set of problems, but accounting for each of them clearly requires at least a tweak and at best a fundamental rewriting of existing models. Their issues go beyond inelegance.

Experimentalists, myself included, have been chasing aesthetically motivated, or partially aesthetically motivated, theories through the data. After a few years of running the Large

Hadron Collider at the energy frontier and a host of careful measurements in particle, nuclear, and atomic physics carried out all over the globe, large regions of "new physics" parameter space have been excluded. Theorists have answered with pivots and extensions, adapting their proposed models in ways that push us to adopt more challenging experimental conditions.

This exchange has seemed healthy and has definitely been fun. The close interactions have allowed for fast progress in testing new ideas. Even though the searches for non–Standard Model physics have resulted in new limits rather than discovery, it has been thrilling to make measurements that might provide evidence toward a Grand Unified Theory. However, our current era of scarce resources requires tighter thinking. It's time to more carefully scrutinize our theoretical foundations.

Including aesthetic considerations in the scientific toolbox has resulted in huge leaps forward. The drive for elegance has repeatedly enabled scientists to uncover underlying structure. The permission to consider aesthetics is part of what drew many of us to become scientists in the first place. I'm not arguing that we ought to abandon it forever. But we're currently in a data-rich period in particle physics, after years of being (at least at the energy frontier) data-poor. Ensuring that data get the final say is more essential than anything else in the practice of science, and the data we have in hand could say a lot about the Standard Model. There's even more on the line when we consider which experiments to pursue next.

At this stage, with 96 percent of the universe's content in the dark, it's a mistake to put aesthetic concerns on a par with contradictions when it comes to theoretical motivation. With no explanation for dark energy, no confirmed detection of dark matter, and no sufficient mechanism for matter/antimatter asymmetry, we have too many gaps to worry about elegance. Theorists will keep pushing for Grand Unified Theories, including developing the mathematics to enable further progress. Experimentalists have an opportunity and a responsibility to provide direction, through agnostic hunts for discrepancies between our data and Standard Model predictions. This includes, of course, measuring the hell out of the newly discovered Higgs boson.

It's time for us to admit that some of the models we've been chasing from our brilliant theory colleagues might actually be (gorgeous) Hail Mary passes to the universe. Our next significant level of understanding will likely come because we're forced there by painstakingly determined constraints from the data rather than by a lucky catch.

NATURALNESS, HIERARCHY, AND SPACETIME

MARIA SPIROPULU

Professor of physics, Caltech

Naturalness, hierarchy, and spacetime, as invoked today in physics, will be retired sooner than later.

The naturalness "strategy" and hierarchy "problem" for building models toward theories that extend the Standard Model of particles and their interactions (call it STh, standard theory à la David Gross) are crumbling with the measurements of the newly discovered Higgs-like boson. I'll call it "Higgs-like" until we've measured it exhaustively at the Large Hadron Collider. Nonetheless, we've built ourselves a story for what comes after the Higgs elementary scalar—a story that the real world doesn't seem to abide by.

So, our enslavement by the need to be "natural," not "finely-tuned," (subjective notions we should have objected to much earlier) is being lifted as we speak, and the road to high energy may be surprisingly more complex than what we were envisioning.

Toward the end of the road (and there may be none such, if the road curves back at us), there is gravity and spacetime entering the mix of physics notions that are hairy and loopy, and we'll have to upgrade them, too, if not retire them altogether.

On related physics ideas, the notions about the particle nature of dark matter might also crumble. Some big revolutions (and discoveries) are in store regarding fundamental notions about our quantum universe.

SCIENTISTS OUGHT TO KNOW EVERYTHING SCIENTIFICALLY KNOWABLE

ED REGIS

Science writer; coauthor (with George Church), Regenesis

In 1993, two Nobel prize–winning physicists, Steven Weinberg and Leon Lederman, each published books suggesting that a 54-mile-long particle accelerator, the Superconducting Super Collider (SSC), should be constructed near Waxahachie, Texas, in order to discover the elusive Higgs scalar boson, which Lederman had semi-facetiously dubbed "the God particle." (The books were *Dreams of a Final Theory*, and *The God Particle*, respectively.) In a *tour de force* of bad timing, both books came out just as the U.S. Congress was in the process of terminating funding for the project once and for all.

Which was just as well. As it happened, the Higgs boson was discovered in 2012 by scientists working at a much smaller accelerator, the 17-mile-long Large Hadron Collider (LHC) at CERN, near Geneva.

As often happens in science, a new discovery simultaneously raises several new questions, which of course was also the case with the Higgs. For instance, Why did the Higgs particle have precisely the mass it had? Were there yet even more basic particles that lay beneath, and explained, certain attributes of the Higgs? Was there in fact more than one Higgs boson? In fundamental particle theory, unfortunately, the answers to such questions have become increasingly, and even prohibitively, expensive. Before it was canceled, cost estimates for the SSC rose from an initial $3.9 billion to a final $11-billion-plus in 1991. But how much is it really worth to know the answers to further questions regarding the Higgs particle? How much, if anything, would *you* pay to know those answers, assuming, optimistically, that you could even understand such questions as, How does the Higgs boson explain (if at all) the phenomenon of electroweak symmetry-breaking?

Science has long since reached the point where some types of new knowledge can be discovered only by building structures so absurdly cosmic, and even comic, in size as to have equally cosmic price tags. In light of this, it makes sense to ask whether the knowledge supposedly to be provided by these dollar-bill-destroying behemoths is in fact worth acquiring.

Apparently unfazed by congressional rejection of the Super Collider, a 2001 study group at Fermilab (whose accelerator was a relatively puny 4 miles around) seriously entertained the prospect of building a *Very* Large Hadron Collider (VLHC), a stupendous monster that would be fully 233 kilometers (145 miles) in circumference. This leviathan object would enclose an area more than 400 square miles larger than the state of Rhode Island.

Then, in the summer of 2013, a year after the Higgs had been discovered, a group of particle physicists met at Minneapolis to propose a 62-mile-long collider that, they said, would allow "the study of indirect effects of new physics on the W and Z bosons, the top quark, and other systems."[d] These proposals just keep coming, like spam, junk mail, or crabgrass. But sooner or later, enough has got to be enough, even in science, which, after all, is not sacrosanct. It's just silly to keep paying—forever, eternally, and in perpetuity—more and more money for less and less knowledge about hypothetical specks of matter that go so far beyond the infinitesimal as to border on sheer nothingness.

Fundamental-particle physicists, evidently, have never heard of "limits to growth," or limits of any other kind. But they should certainly acquaint themselves with that concept, for the fundamental does not automatically trump the practical. Every dollar spent on a shiny new megacollider is a dollar that can't be spent on other things, such as hospitals, vaccine development, epidemic prevention, disaster relief, and so on. Particle accelerators the size of small nations are arguably well over the financial horizon of what's reasonable to sacrifice for a given incremental advance in arcane, theoretical, almost cabalistic knowledge. In a postmortem on the Superconducting Super Collider ("Good-bye to the SSC"), science historian Daniel J. Kevles wrote that basic research in physics should be pursued, "But not at any price."[e] I agree. Some scientific knowledge is simply not worth its cost.

FALSIFIABILITY

SEAN CARROLL
Theoretical physicist, Caltech; author, The Particle at the End of the Universe

In a world where scientific theories often sound bizarre and counter to everyday intuition and a wide variety of nonsense aspires to be recognized as "scientific," it's important to be able to separate science from non-science—what philosophers call "the demarcation problem." Karl Popper famously suggested the criterion of "falsifiability": A theory is scientific if it makes clear predictions that can be unambiguously falsified.

It's a well-meaning idea but far from the complete story. Popper was concerned with theories such as Freudian psychoanalysis and Marxist economics, which he considered non-scientific. No matter what actually happens to people or societies, Popper claimed, theories like these will always be able to tell a story in which the data are compatible with the theoretical framework. He contrasted this with Einstein's relativity, which made specific quantitative predictions ahead of time. (One prediction of general relativity was that the universe should be expanding or contracting, leading Einstein to modify the theory because he thought the universe was actually static. So even in this example, the falsifiability criterion is not as unambiguous as it seems.)

Modern physics stretches into realms far removed from everyday experience, and sometimes the connection to experiment becomes tenuous at best. String theory and other approaches to quantum gravity involve phenomena that are likely to manifest themselves only at energies enormously higher than anything we have access to here on Earth. The cosmological multiverse and the many-worlds interpretation of quantum mechanics posit other realms impossible for us to access directly. Some scientists, leaning on Popper, have suggested that these theories are non-scientific because they're not falsifiable.

The truth is the opposite. Whether or not we can observe them directly, the entities involved in these theories are either real or they are not. Refusing to contemplate their possible existence on the grounds of some *a-priori* principle, even though they might play a crucial role in how the world works, is as non-scientific as it gets.

The falsifiability criterion gestures toward something true and important about science, but it's a blunt instrument in a situation calling for subtlety and precision. It's better to emphasize two central features of good scientific theories: They're *definite* and they're *empirical*. By "definite," we mean that they say something clear and unambiguous about how reality functions. String theory says that in certain regions of parameter space, ordinary particles behave as loops or segments of one-dimensional strings. The relevant parameter space might be inaccessible to us, but it's part of the theory which cannot be avoided. In the cosmological multiverse, regions unlike our own are unambiguously there even if we can't reach them. This is what distinguishes these theories from the approaches Popper was trying to classify as non-scientific. (Popper himself understood that theories should be falsifiable "in principle," but that modifier is often forgotten in contemporary discussions.)

It's the "empirical" criterion that requires some care. On the face of it, this criterion might be mistaken for "makes falsifiable predictions." But in the real world, the interplay between theory and experiment isn't so cut and dried. A scientific theory is ultimately judged by its ability to account for the data—but the steps along the way to that accounting can be indirect.

Consider the multiverse, often invoked as a potential solution to some of the fine-tuning problems of contemporary cosmology. For example, we believe there's a small but nonzero vacuum energy inherent in empty space. This is the leading theory to explain the observed acceleration of the universe, for which the 2011 Nobel Prize in physics was awarded. The problem for theorists is not that vacuum energy is hard to explain; it's that the predicted value is enormously larger than what we observe.

If the universe we see around us is the only one there is, the vacuum energy is a unique constant of nature and we're faced with the problem of explaining it. If, on the other hand, we live in a multiverse, the vacuum energy could be completely different in different regions, and an explanation suggests itself immediately: In regions where the vacuum energy is much larger, conditions are inhospitable to the existence of life. There is therefore a selection effect, and we should predict a small value of the vacuum energy. Indeed, using this precise reasoning, Steven Weinberg did predict the value of the vacuum energy long before the acceleration of the universe was discovered.

We can't (as far as we know) observe other parts of the multiverse directly, but their existence has a dramatic effect on how we account for the data in the part of the multiverse we do observe. It's in that sense that the success or failure of the idea is ultimately empirical: Its virtue is not that it's a neat idea, or fulfills some nebulous principle of reasoning, but that it helps us account for the data. Even if we'll never visit those other universes.

Science isn't merely armchair theorizing, it's about explaining the world we see, developing models that fit the data. But fitting models to data is a complex and multifaceted process, involving a give-and-take between theory and experiment, as well as the gradual development of theoretical understanding in its own right. In complicated situations, fortune-cookie-sized mottos like "Theories should be falsifiable" are no substitute for careful thinking about how science works. Fortunately, science marches on, largely heedless of amateur philosophizing. If string theory and multiverse theories help us understand the world, they'll grow in acceptance. If they prove ultimately too nebulous, or better theories come along, they'll be discarded. The process might be messy, but nature is the ultimate guide.

ANTI-ANECDOTALISM

NICHOLAS G. CARR

Journalist; author, The Glass Cage: Automation and Us

We live anecdotally, proceeding from birth to death through a series of incidents, but scientists can be quick to dismiss the value of anecdotes. "Anecdotal" has become something of a curse word, at least when applied to research and other explorations of the real. A personal story, in this view, is a distraction or a distortion, something that gets in the way of a broader, statistically rigorous analysis of a large set of observations or a big pile of data. But as this year's *Edge* Question makes clear, the line between the objective and the subjective falls short of the Euclidean ideal. It's negotiable. The empirical, if it's to provide anything like a full picture, needs to make room for both the statistical and the anecdotal.

The danger in scorning the anecdotal is that science gets too far removed from the actual experience of life, losing sight of the fact that mathematical averages and other such measures are always abstractions. Some prominent physicists have recently questioned the need for philosophy, implying that it's been rendered obsolete by scientific inquiry. I wonder if that opinion isn't a symptom of anti-anecdotalism. Philosophers, poets, artists—their raw material includes the anecdote, and they remain, even more so than scientists, our best guides to what it means to exist.

SCIENCE MAKES PHILOSOPHY OBSOLETE

REBECCA NEWBERGER GOLDSTEIN

Philosopher, novelist; author, Plato at the Googleplex

The obsolescence of philosophy is often taken to be a consequence of science. After all, science has a history of repeatedly inheriting—and definitively answering—questions over which philosophers have futilely hemmed and hawed for unconscionable amounts of time. It's been that way from the beginning. Those irrepressible ancient Greeks, Thales & Co., in speculating about the ultimate constituents of the physical world and the laws that govern its changes, were asking questions that awaited answers from physics and cosmology. And so it has gone, science transforming philosophy's vagaries into empirically testable theories, right down to our own scientifically explosive period, when the advancement of cognitive and affective neuroscience has brought such questions as the nature of consciousness, free will, and morality—those perennials of the philosophy curriculum—under the gaze of fMRI–enhanced scientists. Philosophy's role in the business of knowledge—or so goes the story—is to send up a signal reading, "Science desperately needed here." Or, changing the metaphor, philosophy is a cold-storage room in which questions are shelved until the sciences get around to handling them. Or, to change the metaphor yet again, philosophers are premature ejaculators who decant too soon, spilling their seminal genius to no effect. Choose your metaphor, the moral of the story is that the history of scientific expansion is the history of philosophical contraction, and the natural progression ends in the elimination of philosophy.

What's wrong with this story? Well, for starters it's internally incoherent. You can't argue for science making philosophy obsolete without indulging in philosophical arguments. You're going to need to argue, for example, for a clear criterion for distinguishing between scientific and nonscientific theories of the world. When pressed for an answer to the so-called demarcation problem, scientists almost automatically reach for the notion of "falsifiability" first proposed by Karl Popper. His profession? Philosophy. But whatever criterion you offer, its defense is going to implicate you in philosophy. Likewise with the unavoidable question—especially for those who argue philosophy's obsolescence—of what it is we're doing in doing science. Are we offering descriptions of reality and so extending our ontology in discovering the entities and forces utilized in our best scientific theories? Have we learned, as scientific realism would have it, that there are genes and neurons, fermions and bosons, perhaps a multiverse? Or are these theoretical terms not meant to be interpreted as references to things in the world at all but as mere metaphorical gears in the instruments of prediction known as theories? Presumably scientists care about the philosophical question of whether they're actually talking about anything other than observations when they do their science. Even more to the point, the view that science eliminates philosophy requires a philosophical defense of scientific realism. (And if you think not, then that's going to require a philosophical argument.)

A triumphalist scientism needs philosophy to support itself. And the lesson here should be generalized. Philosophy is joined to science in reason's project. Its mandate is to render our views and our attitudes maximally coherent. This involves it in the task of (in Wilfrid Sellars' terms) reconciling the "scientific" and the "manifest" images we have of our being in the world, which also involves philosophy in providing the reasoning that science requires in order to claim its image as descriptive.

Perhaps the old demarcation problem of distinguishing the scientific is misguided. The more important demarcation is distinguishing all that is implicated in and reconcilable with the scientific claims of knowledge. This leads me to hazard a more utopian answer to this year's *Edge* Question than the one I proposed in the title. What idea should science retire? The idea of "science" itself. Let's retire it in favor of the more inclusive "knowledge."

"SCIENCE"

IAN BOGOST

Video game designer; Ivan Allen College Distinguished Chair in Media Studies and professor of interactive computing, Georgia Institute of Technology; author, Alien Phenomenology

"No topic is left unexplored," reads the jacket blurb of *The Science of Orgasm*, a 2006 book by an endocrinologist, a neuroscientist, and a "sexologist." A list of topics covered includes the genital-brain connection and how the brain produces orgasms. The result, promises the jacket blurb, "illuminates the hows, whats, and wherefores of orgasm."

Its virtues or faults notwithstanding, *The Science of Orgasm* exemplifies a trend that has become nearly ubiquitous in popular discourse—that a topic can be best and most thoroughly understood from the vantage point of "science." How common is this approach? Google Books produces nearly 150 million search results for the phrase "the science of"—including dozens of books with the quip in their titles. The science of smarter spending; the science of acting; the science of champagne; the science of fear; the science of composting—the list goes on.

"The science of *X*" is one example of the rhetoric of science—the idea that anything called "science" is science—but not the only one. There's also "scientists have shown" or its commoner shorthand, "studies show"—phrases that appeal to the authority of science whether or not the conclusions they summarize bear any resemblance to the purported studies from which those conclusions were derived.

Both of these tendencies could rightly be accused of *sci-*

entism, the view that empirical science entails the most complete, authoritative, and valid approach to answering questions about the world. Scientism isn't a new erroneous notion but it's an increasingly popular one. Recently, Stephen Hawking pronounced philosophy "dead" because it hasn't kept up with advances in physics. Scientism assumes that the only productive way to understand the universe is through the pursuit of science and that all other activities are lesser at best, pointless at worst.

And to be sure, the rhetoric of science has arisen partly thanks to scientism. "Science of *X*" books, and research findings traceable to an origin in apparently scientific experimentation, increasingly take the place of philosophical, interpretive, and reflective accounts of the meaning and importance of activities of all kinds. Instead of pondering the social uses of sparkling wine and its pleasures, we ponder what the size of its bubbles indicates about its quality, or why that effervescence lasts longer in a modern fluted glass than in a wider champagne coupe.

But the rhetoric of science doesn't just risk the descent into scientism. It also gives science sole credit for something it doesn't deserve credit for: an attention to the construction and operation of things. Most of the "science of *X*" books look at the material form of their subject, be it neurochemical, computational, or economic. But attention to the material realities of a subject has no necessary relationship to science at all. Literary scholars study the history of the book, including its material evolution from clay tablet to papyrus to codex. Artists rely on a deep understanding of the physical mediums of pigment, marble, or optics when they fashion creations. Chefs require a sophisticated grasp of the chemistry and biology of food in order to thrive in their craft. To think that science has a special relationship to observations about the material world isn't just wrong, it's insulting.

Beyond encouraging people to see science as the only direction for human knowledge and absconding with the subject of materiality, the rhetoric of science also does a disservice to science itself. It makes science look simple, easy, and fun, when it's mostly complex, difficult, and monotonous.

A case in point: The popular Facebook page "I f★cking love science" posts quick-take variations on the "science of *X*" theme—mostly images and short descriptions of unfamiliar creatures, like the pink fairy armadillo, or illustrated birthday wishes to famous scientists like Hawking. But as the science fiction writer John Skylar rightly insisted in a fiery takedown of the practice last year, most people don't f★cking love science, they f★cking love photography—pretty images of fairy armadillos and renowned physicists. The pleasure derived from these pictures obviates the public's need to understand how science actually gets done—slowly and methodically, with little acknowledgment and modest pay in unseen laboratories and research facilities.

The rhetoric of science has consequences. Things that have no particular relation to scientific practice must increasingly frame their work in scientific terms to earn any attention or support. The sociology of Internet use suddenly transformed into "Web science." Long-accepted practices of statistical analysis have become "data science." Thanks to shifting educational and research funding priorities, anyone who can't claim membership in a STEM (science, technology, engineering, and math) field will be left out in the cold. Unfortunately, the rhetoric of science offers the tactical response to such new challenges. Unless humanists reframe their work as "literary science," they risk being marginalized, unfunded, and forgotten.

When you're selling ideas, you have to sell the ideas that will sell. But in a secular age in which the abstraction of "science" risks replacing all other abstractions, a watered-down, bland, homogeneous version of science is all that will remain if the rhetoric of science is allowed to prosper.

We need not choose between God and man, science and philosophy, interpretation and evidence. But ironically, in its quest to prove itself as the supreme form of secular knowledge, science has inadvertently elevated itself into a theology. Science is not a practice so much as an ideology. We don't need to destroy science in order to bring it down to earth. But we do need to bring it down to earth again, and the first step in doing so is to abandon the rhetoric that has become its most popular devotional practice.

OUR NARROW DEFINITION OF "SCIENCE"

SAM HARRIS

Neuroscientist; cofounder and chairman, Project Reason; author, Waking Up: A Guide to Spirituality Without Religions

Search your mind, or pay attention to the conversations you have with other people, and you'll discover that there are no real boundaries between science and philosophy—or between those disciplines and any other that attempts to make valid claims about the world on the basis of evidence and logic. When such claims and their methods of verification admit of experiment and/or mathematical description, we tend to say our concerns are "scientific"; when they relate to matters more abstract, or to the consistency of our thinking itself, we often say we're being "philosophical"; when we merely want to know how people behaved in the past, we dub our interests "historical" or "journalistic"; and when a person's commitment to evidence and logic grows dangerously thin or simply snaps under the burden of fear, wishful thinking, tribalism, or ecstasy, we recognize that he's being "religious."

The boundaries between true intellectual disciplines are currently enforced by little more than university budgets and architecture. Is the Shroud of Turin a medieval forgery? This is a question of history, of course, and of archaeology, but the techniques of radiocarbon dating make it a question of chemistry and physics as well. The real distinction we should care about—the observation of which is the *sine qua non* of the scientific attitude—is between demanding good reasons for what one believes and being satisfied with bad ones.

The scientific attitude can handle whatever happens to be the case. Indeed, if the evidence for the inerrancy of the Bible and the resurrection of Jesus Christ were *good*, one could embrace the doctrine of fundamentalist Christianity *scientifically*. The problem, of course, is that the evidence is either terrible or nonexistent—hence the partition we have erected (in practice, never in principle) between science and religion.

Confusion on this point has spawned many strange ideas about the nature of human knowledge and the limits of "science." People who fear the encroachment of the scientific attitude—especially those who insist upon the dignity of believing in one or another Iron Age god—will often make derogatory use of words such as *materialism, neo–Darwinism,* and *reductionism*, as if those doctrines had some necessary connection to science itself.

There are, of course, good reasons for scientists to be materialist, neo–Darwinian, and reductionist. However, science entails none of those commitments, nor do they entail one another. If there were evidence for dualism (immaterial souls, reincarnation), one could be a scientist without being a materialist. As it happens, the evidence here is extraordinarily thin, so virtually all scientists are materialists of some sort. If there were evidence against evolution by natural selection, one could be a scientific materialist without being a neo–Darwinist. But as it happens, the general framework put forward by Darwin is as well established as any other in science. If there were evidence that complex systems produced phenomena that cannot be understood in terms of their constituent parts, it would be possible to be a neo–Darwinist without being a reductionist. For all practical purposes, that's where most scientists find themselves, because every branch of science beyond physics must resort to concepts that cannot be understood merely in terms of particles and fields. Many of us have had "philosophical" debates about what to make of this explanatory impasse. Does the fact that we cannot predict the behavior of chickens or fledgling democracies on the basis of quantum mechanics mean that those higher-level phenomena are something *other* than their underlying physics? I would vote "no" here, but that doesn't mean I envision a time when we'll use only the nouns and verbs of physics to describe the world.

But even if one thinks the human mind is entirely the product of physics, the reality of consciousness becomes no less wondrous, and the difference between happiness and suffering no less important. Nor does such a view suggest that we'll ever find the emergence of mind from matter fully intelligible; consciousness may always seem like a miracle. In philosophical circles, this is known as "the hard problem of consciousness"—some of us agree that this problem exists, some of us don't. Should consciousness prove conceptually irreducible, remaining the mysterious ground for all we can conceivably experience or value, the rest of the scientific worldview would remain perfectly intact.

The remedy for all this confusion is simple: We must abandon the idea that science is distinct from the rest of human rationality. When you are adhering to the highest standards of logic and evidence, you are thinking scientifically. And when you're not, you're not.

THE HARD PROBLEM

DANIEL C. DENNETT

Philosopher; Austin B. Fletcher Professor of Philosophy, codirector, Center for Cognitive Studies, Tufts University; coauthor (with Linda LaScola), Caught in the Pulpit: Leaving Belief Behind

One might object that the hard problem of consciousness (so dubbed by philosopher David Chalmers in his 1996 book, *The Conscious Mind*) isn't a scientific idea at all, and hence isn't an eligible candidate for this year's Edge Question, but since the philosophers who have adopted the term have also persuaded quite a few cognitive scientists that their best scientific work addresses only the "easy" problems of consciousness, this idea qualifies as scientific: It constrains scientific thinking, distorting scientists' imaginations as they attempt to formulate genuinely scientific theories of consciousness. (I won't give examples, since we're instructed to go after ideas, not people, in our answers.)

No doubt on first acquaintance the philosophers' thought experiments succeed handsomely at pumping the intuitions that zombies are "conceivable" and hence "possible" and that this prospect, the (mere, logical) possibility of zombies, "shows" that there's a hard problem of consciousness untouched by any neuroscientific theories of how consciousness modulates behavioral control, introspective report, emotional responses, etc., etc. But if the scientists impressed by this "result" from philosophers were to take a good hard look at the critical literature in philosophy exploring the flaws in these thought experiments, they would—I hope—recoil in disbelief. (I am embarrassed by the mere thought of them wading through our literature on these topics.)

You see, the arguments implicit in the simple, first-pass thought experiments don't go through without some shoring up. We have to define not just conceivability but ideal conceivability, and then ideal positive conceivability (as distinct from ideal negative conceivability, etc., etc.). Are perpetual motion machines imaginable but ideally inconceivable, or ideally positively conceivable? It makes a big difference, one is told, whether one can "modally imagine" a zombie. What can *you* modally imagine, and are you sure? And Frank Jackson's intuition pump about Mary, the color scientist who's prevented from seeing colors, has to be embellished with imaginary gadgets that prevent her from dreaming in color. Or perhaps she's born color-blind (but otherwise with an entirely normal brain!), or perhaps she's fitted with locked-on goggles displaying black-and-white TV to her poor eyeballs. And that's just a fraction of the complicated fantasies that have been earnestly proposed and rebutted.

I'm not recommending that scientists do this homework, but if they're curious to see what contortions philosophers will inflict upon themselves in order to "save" these retrograde intuitions, they could consult the superhumanly patient analysis of Amber Ross of the University of North Carolina, who dismantles the whole tangled mess in her 2013 PhD dissertation, "Inconceivable Minds."

Is the hard problem an idea that demonstrates the need for a major revolution in science if consciousness is ever to be explained, or an idea that demonstrates the frailties of human imagination? That question is not settled at this time, so scientists should consider adopting the cautious course that postpones all accommodation with it. That's how most neuroscientists handle ESP and psychokinesis—assuming, defeasibly, that they are figments of imagination.

THE NEURAL CORRELATES OF CONSCIOUSNESS

SUSAN BLACKMORE

Psychologist; author, Consciousness: An Introduction

Consciousness is a hot topic in neuroscience, and some of the brightest researchers are hunting for the neural correlates of consciousness (NCCs)—but they'll never find them. The implicit theory of consciousness underlying this quest is misguided and needs to be retired.

The idea of the NCCs is simple enough and intuitively tempting. If we believe in the "hard problem of consciousness"—the mystery of how subjective experience arises from (or is created by or generated by) objective events in a brain—then it's easy to imagine that there must be a special place in the brain where this happens. Or if there's no special place, then some kind of "consciousness neuron," or process or pattern or series of connections. We may not have the first clue how any of these objective things could produce subjective experience, but if we could identify which of them was responsible (so the thinking goes), we'd be one step closer to solving the mystery.

This sounds eminently sensible, as it means taking the well-worn scientific route of starting with correlations before moving on to causal explanations. The trouble is, it depends on a dualist—and ultimately unworkable—theory of consciousness. The underlying intuition is that consciousness is an extra—something additional to and different from the physical processes on which it depends. Searching for the NCCs relies on this difference. On one side of the correlation, you measure neural processes, using EEG, fMRI, or other kinds of brain scan; on the other, you measure subjective experiences, or "consciousness itself." But how?

A popular method is to use binocular rivalry, or ambiguous figures that can be seen in either of two incompatible ways, such as a Necker cube that flips between two orientations. To find the NCCs, you find out which version is being consciously perceived as the perception flips from one to the other, and then you correlate that with what's happening in the visual system. The problem is that the person has to tell you in words, "Now I am conscious of this" or "Now I'm now conscious of that." They might instead press a lever or button (and other animals can do this, too), but in every case you're measuring physical responses.

Is this capturing something called consciousness? Will it help us solve the mystery? No.

This method is really no different from any other correlational studies of brain function, such as correlating activity in the fusiform face area with seeing faces, or in the prefrontal cortex with certain kinds of decision making. It correlates one type of physical measure with another. This isn't useless research. It's interesting to know, for example, where in the visual system neural activity changes when the reported visual experience flips. But discovering this doesn't tell us that this neural activity is the generator of something special called "consciousness" or "subjective experience" while everything else going on in the brain is "unconscious."

I can understand the temptation to think it is. Dualist thinking comes naturally to us. We feel as though our conscious experiences were of a different order from the physical world. But this is the same intuition that leads to the hard problem seeming hard. It's the same intuition that produces the philosopher's zombie—a creature identical to me in every way except that it has no consciousness. It's the same intuition that leads people to write, apparently unproblematically, about brain processes being either conscious or unconscious.

Am I really denying this difference? Yes. Intuitively plausible as it is, this is a magic difference. Consciousness is not some weird and wonderful product of some brain processes but not others. Rather, it's an illusion constructed by a clever brain and body in a complex social world. We can speak, think, refer to ourselves as agents, and so build up the false idea of a persisting self that has consciousness and free will.

We're tricked by an odd feature of consciousness. When I ask myself, "What am I conscious of now?" I can always find an answer. It's the trees outside the window, the sound of the wind, the problem I'm worried about and cannot solve—or whatever seems most vivid at the time. This is what I mean by being conscious now, by having qualia. But what was happening a moment before I asked? When I look back, I can use memories to claim I was conscious of this or that and not conscious of something else, relying on the clarity, logic, consistency, and other such features to decide.

This leads all too easily to the idea that while people are awake they must always be conscious of something or other. And that leads along the slippery path to the idea that if we knew what to look for, we could peer inside someone's brain and find out which processes were the conscious ones and which the unconscious ones. But this is all nonsense. All we'll ever find are the neural correlates of thoughts, perceptions, memories, and the verbal and attentional processes that lead us to think we're conscious.

When we finally have a better theory of consciousness to replace these popular delusions, we'll see that there's no hard problem, no magic difference, and no NCCs.

LONG-TERM MEMORY IS IMMUTABLE

TODD C. SACKTOR

Professor of physiology and pharmacology, professor of neurology, State University of New York Downstate Medical Center

For over a century, psychological theory held that once memories are consolidated from short-term into long-term form, they remain stable and unchanging. Whether certain long-term memories are slowly forgotten or are always present but cannot be retrieved was a matter of debate.

For the last fifty years, research on the neurobiological basis of memory seemed to support the psychological theory. Short-term memory was found to be mediated by biochemical changes at synapses, modifying their strength. Long-term memory was strongly correlated with long-term changes in the number of synapses, either increases or decreases. This intuitively made sense. Biochemical changes were rapid and could be quickly reversed, just like short-term memories. On the other hand, synapses, although small, were anatomical structures visible under the microscope and thus were thought to be stable for weeks, perhaps for years. Short-term memories could easily be prevented from consolidating into the long term by dozens of inhibitors of different signaling molecules. In contrast, there was no known agent that erased a long-term memory.

Two recent lines of evidence show this dominant theory of long-term memory to be ready for retirement. The first is the discovery of reconsolidation. When memories are recalled, they undergo a brief period in which they're once again susceptible to disruption by many of the same biochemical inhibitors that affect the initial conversion of short- into long-term memory. This means that long-term memories are not immutable but can be converted into short-term memory and then reconverted back into long-term memory. If this reconversion doesn't happen, the specific long-term memory is effectively disrupted.

The second is the discovery of a few agents that do indeed erase long-term memories. These include inhibitors of the persistently active enzyme PKMzeta and of a protein translation factor with prionlike properties of perpetuation. Conversely, increasing the activity of the molecules enhances old memories. The persistent changes in synapse number that so strongly correlate with long-term memory may therefore be downstream of persistent biochemical changes. That memory-erasing agents are so few suggests that there may be a relatively simple mechanism for long-term memory storage involving not hundreds of molecules, as in short-term memory, but only a handful, perhaps working together.

Memory reconsolidation allows specific long-term memories to be manipulated. Memory erasure is extraordinarily potent and likely disrupts many if not all long-term memories at the same time. When these two fields are combined, specific long-term memories will be erased or strengthened in ways never conceivable in prior theories.

THE SELF

BRUCE HOOD

Professor of Developmental Psychology in Society, School of Experimental Psychology, University of Bristol, U.K.; author, The Self Illusion

It seems almost redundant to call for the retirement of the free willing self, as the idea is neither scientific nor is this the first time the concept has been dismissed for lack of empirical support. The self did not have to be discovered; it's the default assumption most of us experience, so it wasn't really revealed by methods of scientific inquiry. Challenging the notion of a self is also not new. Freud's unconscious ego has been dismissed for lack of empirical support since the cognitive revolution of the 1950s.

Yet the self, like a conceptual zombie, refuses to die. It crops up again and again in recent theories of decision making, as an entity with free will which can be depleted. It reappears as an interpreter in cognitive neuroscience, as able to integrate parallel streams of information arising from separable neural substrates. Even if these appearances of the self are understood to be convenient ways of discussing the emergent output of multiple parallel processes, students of the mind continue to implicitly endorse the idea that there's a decision maker, an experiencer, a point of origin.

We know the self is constructed because it can be so easily deconstructed—through damage, disease, and drugs. It must be an emergent property of a parallel system processing input, output, and internal representations. It's an illusion because it feels so real, but that experience is not what it seems. The same is true for free will. Although we can experience the mental anguish of making a decision, our free will cannot be some kind of King Solomon in our minds, weighing the pros and cons, as this would present the problem of logical infinite regress (who is inside *his* head, and so on?). The choices and decisions we make are based on situations that impose on us. We don't have the free will to choose the experiences that have shaped our decisions.

Should we really care about the self? After all, trying to live without the self is challenging and not how we think. By experiencing, evoking, and talking about the self, we're conveniently addressing a phenomenology we can all relate to. Defaulting to the self in explanations of human behavior enables us to draw an abrupt stop in the chain of causality when trying to understand thoughts and actions. How notable that one does this so easily when talking about humans but when one applies the same approach to animals one is accused of anthropomorphism!

By abandoning the free willing self, we're forced to reexamine the factors that are truly behind our thoughts and behavior and the way they interact, balance, override, and cancel out. Only then will we begin to make progress in understanding how we really operate.

COGNITIVE AGENCY

THOMAS METZINGER
Philosopher, Johannes Gutenberg-Universität Mainz; author, The Ego Tunnel

Thinking isn't something you do. Most of the time, it's something that happens to you. Cutting-edge research on the phenomenon of Mind Wandering clearly shows how almost all of us, for more than two-thirds of our conscious lifetime, are not in control of our conscious thought processes.

Western culture, traditional philosophy of mind, and even cognitive neuroscience have been deeply influenced by the Myth of Cognitive Agency. It's the myth of the Cartesian Ego, the active thinker of thoughts, the epistemic subject that acts—mentally, rationally, in a goal-directed manner—and can always terminate or suspend its own cognitive processing at will. It's the theory that conscious thought is a personal-level process—something that by necessity has to be ascribed to you, the person as a whole. This theory has now been empirically refuted. As it turns out, most of our conscious thoughts are actually the product of subpersonal processes, like breathing or the peristaltic movements in our gastrointestinal tract. The Myth of Cognitive Agency says that we're mentally autonomous beings. We can now see that this is an old complacent fairy tale. It's time to put it to rest.

Recent studies in the booming research field of Mind Wandering show that we spend roughly two-thirds of our conscious lifetime zoning out—daydreaming, lost in fantasies, autobiographical planning, inner narratives, or depressive rumination. Depending on the study, 30 to 50 percent of our waking life is occupied by spontaneously occurring stimulus and task-unrelated thought. Mind Wandering probably has positive aspects, too, because it's associated with creativity, careful future planning, or the encoding of long-term memories. But its overall performance costs (for example, in terms of reading comprehension, memory, sustained-attention tasks, or working memory) are marked and have been well documented. So have its negative effects on general, subjective well-being. A wandering mind is an unhappy mind, but it may only be part of a more comprehensive process beyond the conscious self's control or understanding. The sudden loss of inner autonomy—which all of us experience many hundreds of times every day—seems to be based on a cyclically recurring process in the brain. The ebb and flow of autonomy and meta-awareness might well be a kind of attentional seesawing between our inner and outer worlds, caused by a constant competition between the brain networks underlying spontaneous subpersonal thinking and goal-oriented cognition.

Mind Wandering is not the only way in which our attention gets decoupled from perception of the Here and Now. There are also periods of "mind blanking"; these episodes often may not be remembered and also frequently escape detection by external observers. In addition, there's clearly complex but uncontrollable cognitive phenomenology during sleep. Adults spend approximately one-and-a-half to two hours per night in REM sleep, experiencing dreams in which they're mostly unable to control their conscious thought process. NREM [non–rapid eye movement] sleep yields similar, dreamlike reports during stage 1, whereas other stages of NREM sleep are characterized by mostly cognitive/symbolic mentation that is typically confused, nonprogressive, and perseverative. A conservative estimate would therefore be that for much more than half our lifetime we're not cognitive agents in the true sense of the word. This excludes periods of illness, intoxication, or insomnia, in which people suffer from dysfunctional forms of cognitive control such as thought suppression, worry, rumination, and counterfactual imagery and are plagued by intrusive thoughts, feelings of regret, shame, and guilt. We don't yet know when and how children acquire a conscious self-model that permits controlled, rational thought. But another sad yet empirically plausible assumption is that most of us gradually lose cognitive autonomy toward the end of life.

Interestingly, the neural correlate of non-autonomous conscious thought overlaps to a considerable degree with ongoing activity in what neuroscientists call the "default mode network." One global function of Mind Wandering might be called "autobiographical self-model maintenance." Mind Wandering creates an adaptive form of self-deception—namely, an illusion of personal identity across time. It helps maintain a fictional "self" that then lays the foundation for important achievements like reward prediction or delay discounting. As a philosopher, my conceptual point is that only if an organism simulates itself as being one and the same across time will it be able to represent reward events or the achievement of goals as a fulfillment of its own goals—as happening to the same entity. I like to call this "the principle of virtual identity formation." Many higher forms of intelligence and adaptive behavior, including risk management, moral cognition, and cooperative social behavior, functionally presuppose a self-model that portrays the organism as a single entity enduring over time. Because we are really only cognitive systems (that is, complex processes without any precise identity criteria) the formation of an (illusory) identity across time can be achieved only on a virtual level—for example, through the creation of an automatic narrative. This could be the more fundamental and overarching computational goal of Mind Wandering and one it may share with dreaming. If I'm right, the default mode of the autobiographical self-modeling constructs a domain-general functional platform enabling long-term motivation and future planning.

Mental autonomy, and how it can be improved, will be one of the hottest topics for the future. There's even a deep link between mental and political autonomy—you cannot sustain one without the other. Because there are not only bodily actions but also mental actions, autonomy has to do with freedom, in one of the deepest and most fundamental senses of the word. But the ability to act autonomously implies more than reasons, arguments, and rationality. Much more fundamentally, it refers to the ability to willfully inhibit, suspend, or terminate our own actions—bodily, social, or mental. The breakdown of this ability is what we call Mind Wandering. It's not an inner action at all but a form of unintentional behavior, an involuntary form of mental activity.

FREE WILL

JERRY COYNE

Professor, Department of Ecology and Evolution, University of Chicago; author, Why Evolution Is True

Among virtually all scientists, dualism is dead. Our thoughts and actions are the outputs of a computer made of meat, our brain—a computer that must obey the laws of physics. Our choices, therefore, must also obey those laws. This puts paid to the traditional idea of dualistic or "libertarian" free will: that our lives comprise a series of decisions in which we could have chosen otherwise. We know now that we can never do otherwise, and we know it in two ways.

The first is from scientific experience, which shows no evidence for a mind separate from the physical brain. This means that "I"—whatever "I" means—may have the illusion of choosing, but my choices are in principle predictable by the laws of physics (excepting any quantum indeterminacy that acts in my neurons). In short, the traditional notion of free will—defined by the biologist Anthony Cashmore as "a belief that there is a component to biological behavior that is something more than the unavoidable consequences of the genetic and environmental history of the individual and the possible stochastic laws of nature"[f]—is dead on arrival.

Second, recent experiments support the idea that our "decisions" often precede our consciousness of having made them. Increasingly sophisticated studies using brain scanning show that those scans can often predict the choices one will make several seconds before the subject is conscious of having chosen. Indeed, our feeling of "making a choice" may itself be a post-hoc confabulation, perhaps an evolved one.

When pressed, nearly all scientists and most philosophers admit this. Determinism and materialism, they agree, win the day. But they're remarkably quiet about it. Instead of spreading the important scientific message that our behaviors are the deterministic results of a physical process, they'd rather invent new "compatibilist" versions of free will—versions that comport with determinism. "Well, when we order strawberry ice cream, we really couldn't have ordered vanilla," they say, "but we still have free will in another sense. And it's the only sense that's important."

Unfortunately, what's "important" differs among philosophers. Some say that what's important is that our complex brain evolved to absorb many inputs and run them through complex programs ("ruminations") before giving an output ("decision"). Others say that what's important is that it's our *own* brain and nobody else's that makes our decisions, even if those decisions are predetermined. Some even argue that we have free will because most of us choose without duress: Nobody holds a gun to our head and says "Order the strawberry." But of course that's not true: The guns are the electrical signals in our brain.

In the end, there's nothing "free" about compatibilist free will. It's a semantic game, in which choice becomes an illusion—something that isn't what it seems. Whether or not we can "choose" is a matter for science, not philosophy, and science tells us that we're complex marionettes dancing to the strings of our genes and environments. Philosophy, watching the show, says, "Pay attention to *me*, for I've changed the game."

So why does the term "free will" still hang around when science has destroyed its conventional meaning? Some compatibilists, perhaps, are impressed by their feeling that they *can* choose, and must comport this with science. Others have said explicitly that characterizing "free will" as an illusion will hurt society. If people believe they're puppets, well, then maybe they'll be crippled by nihilism, lacking the will to leave their beds. This attitude reminds me of the (probably apocryphal) statement of the Bishop of Worcester's wife when she heard about Darwin's theory: "My dear, descended from the apes! Let us hope it is not true, but if it is, let us pray it will not become generally known."

What puzzles me is why compatibilists spend so much time trying to harmonize determinism with a historically nondeterministic concept instead of tackling the harder but more important task of selling the public on the scientific notions of materialism, naturalism, and their consequence: The mind is produced by the brain.

These consequences of incompatibilism mean a complete rethinking of how we punish and reward people. When we realize that the person who kills because of a mental disorder had precisely as much "choice" as someone who murders from childhood abuse or a bad environment, we'll see that everyone deserves the mitigation now given only to those deemed unable to choose between right and wrong. For if our actions are predetermined, none of us can make that choice. Punishment for crimes will still be needed, of course, to deter others, rehabilitate offenders, and remove criminals from society. But now this can be put on a more scientific footing: What interventions can best help both society and the offender? And we lose the useless idea of justice as retribution.

Accepting incompatibilism also dissolves the notion of moral responsibility. Yes, we're responsible for our actions, but only in the sense that they're committed by an identifiable individual. But if you can't really choose to be good or bad—to punch someone or save a drowning child—what do we mean by *moral* responsibility? Some may argue that getting rid of that idea also jettisons an important social good. I claim the opposite: By rejecting moral responsibility, we're free to judge actions not by some dictate, divine or otherwise, but by their consequences—what's good or bad for society.

Finally, rejecting free will means rejecting the fundamental tenets of the many religions that depend on freely choosing a god or a savior.

The fears motivating some compatibilists—that a version of free will must be maintained lest society collapse—won't be realized. The illusion of agency is so powerful that even strong incompatibilists like myself will always act as if we had choices, even though we know we don't. We have no choice in this matter. But we can at least ponder why evolution might have bequeathed us such a powerful illusion.

COMMON SENSE

ROBERT PROVINE

Neuroscientist, emeritus professor of psychology, University of Maryland, Baltimore County; author, Curious Behavior

We fancy ourselves intelligent, conscious, and alert, thinking our way through life. This is an illusion. We're deluded by our brain's generation of a sketchy, rational narrative of subconscious, sometimes irrational, or fictitious events that we accept as reality. These narratives are so compelling that they become common sense, and we use them to guide our lives. In cases of brain damage, neurologists use the term "confabulation" to describe a patient's game but flawed attempt to produce an accurate narrative of life events. I suggest we be equally wary of everyday, nonpathological confabulation and retire the commonsense hypothesis that we're rational beings in full conscious control of our lives. Indeed, we may be passengers in our body, just going along for the ride and privy only to secondhand knowledge of our status, course, and destination.

Behavioral and brain science detect chinks in our synthetic, neurologically generated edifice of reality. Research on sensory illusions indicates that percepts are simply our best estimate of the nature of physical stimuli, not a precise rendering of things and events. The image of our own body is an oddly shaped product of brain function. Memory of things past is also fraught with uncertainty; it's not the reading-out of information from the brain's neurological data bank but an ongoing construct, subject to error and bias. The brain also makes decisions and initiates action before the observer is consciously aware of detecting and responding to stimuli. My own research found that people confabulate narratives to rationalize their laughter, such as "It was funny," or "I was embarrassed," neglecting laughter's involuntary nature and frequent contagiousness.

Our lives are guided by a series of these guesstimates about the behavior and mental state of ourselves and others—guesstimates that, although imperfect, are adaptive and sufficiently accurate to enable us to muddle along. However, as scientists we demand more than default explanations based on common sense. Behavioral and brain science provide a path to understanding that challenges the myths of mental life and everyday behavior. One of its delights is that reality is often turned on its head, revealing hidden processes and providing revelations about who we are, what we're doing, and where we're going.

THERE CAN BE NO SCIENCE OF ART

JONATHAN GOTTSCHALL

Distinguished Research Fellow, English Department, Washington & Jefferson College; author, The Storytelling Animal

Fifteen thousand years ago in France, a sculptor swam and slithered almost a kilometer down into a mountain cave. Using clay, the artist shaped a big bull rearing to mount a cow and then left his creation in the bowels of the earth. The two bison of the Tuc D'Audoubert caves sat undisturbed until they were discovered by spelunking boys in 1912—one of many shocking 20th-century discoveries of sophisticated cave art stretching back tens of thousands of years. The discoveries overturned our sense of what our caveman ancestors were like. They weren't furry, grunting troglodytes; they had artistic souls. They showed us that humans are—by nature, not just by culture—art-making, art-consuming, art-addicted apes.

But why? Why did the sculptor burrow into the earth, make art, and leave it there in the dark? And why does art exist in the first place? Scholars have spun a lot of stories in answer to such questions, but the truth is that we don't know. Here's one reason: Science is lying down on the job.

A long time ago, someone proclaimed that art couldn't be studied scientifically, and for some reason almost everyone believed it. The humanities and sciences constituted, as Stephen Jay Gould might have proclaimed, separate, nonoverlapping magisteria, the tools of one radically unsuited to the other.

Science has mostly bought into this. How else can we

explain its neglect of the arts? People live in art. We read stories and watch them on TV and listen to them in song. We make paintings and gaze at them. We beautify our homes like bowerbirds adorning nests. We demand beauty in the products we buy, which explains the gleam of our automobiles and the sleek modernist aesthetic of our iPhones. We make art out of our own bodies, sculpting them through diet and exercise, festooning them with jewelry and colorful garments, using our skin as living canvas for the display of tattoos. So it is, the world over. As the late Denis Dutton argued in *The Art Instinct*, underneath the cultural variations, "all human beings have essentially the same art."

Our curious love affair with art sets our species apart as much as our sapience or our language or our use of tools. And yet we understand so little about art. We don't know why it exists in the first place. We don't know why we crave beauty. We don't know how art produces its effects in our brains—why one arrangement of sound or color pleases, while another cloys. We don't know much about the precursors of art in other species, and we don't know when humans became creatures of art. (According to one influential theory, art arrived 50,000 years ago in a kind of creative Big Bang. If that's true, how did it happen?) We don't even have a good definition, in truth, for what art *is*. In short, there's nothing so central to human life that's so incompletely understood.

Recent years have seen more use of scientific tools and methods in the humanities. Neuroscientists can show us what's happening in the brain when we enjoy a song or study a painting. Psychologists are studying how novels and TV shows shape our politics or our morality. Evolutionary psychologists and literary scholars are teaming up to explore narrative's Darwinian origins. Other literary scholars are developing a "digital humanities," using algorithms to extract Big Data from digitized literature. But scientific work in the humanities has mainly been scattered, preliminary, and desultory. It doesn't constitute a research program.

If we want better answers to fundamental questions about art, science must jump into the game with both feet. Going it alone, humanities scholars can tell intriguing stories about the origins and significance of art, but they don't have the tools to patiently winnow the field of competing ideas. That's what the scientific method is for—separating the more accurate stories from the less accurate stories. But a strong science of art will require both the thick, granular expertise of humanities scholars and the clever hypothesis-testing of scientists. I'm not calling for a scientific takeover of the arts, I'm calling for a partnership.

This partnership faces great obstacles. There's the unexamined assumption that something about art makes it science-proof. There's the widespread if usually unspoken belief that art is just a frill in human life—relatively unimportant compared to the weighty stuff of science. And there's the weird idea that science necessarily destroys the beauty it seeks to explain (as though a learned astronomer could dull the starshine). But the Delphic admonition "Know thyself" still rings out as the great prime directive of intellectual inquiry, and there will always be a gaping hole in human self-knowledge until we develop a science of art.

SCIENCE *AND* TECHNOLOGY

GEORGE DYSON

Science historian; author, Turing's Cathedral: The Origins of the Digital Universe

The phrase "science and technology" presumes an inseparability that may not be as secure as we think. There can be science without technology, and there can be technology without science.

Pure mathematics is one example—from the Pythagoreans to Japanese temple geometry—of a science flourishing without technology. Imperial China developed sophisticated technologies while neglecting science, and it's all too easy to imagine a society that embraces technology but represses science until only technology remains. Or one particular species of technology might achieve such dominance that it halts the advance of science in order to preserve itself.

That science has brought us technology does not mean that technology will always bring us science. Science could go into retirement at any time. Retiring the assumption that as long as technology flourishes, so will science, might help us avoid this mistake.

THINGS ARE EITHER TRUE OR FALSE

ALAN ALDA

Actor, writer, director; host of the PBS program Brains on Trial; *author,* Things I Overheard While Talking to Myself

The idea that things are either true or false should possibly take a rest.

I'm not a scientist, just a lover of science, so I might be speaking out of turn—but like all lovers I think about my beloved a lot. I want her to be free and productive, and not misunderstood.

For me, the trouble with truth is that not only is the notion of eternal, universal truth highly questionable, but simple, local truths are subject to refinement as well. Up is up and down is down, of course. Except under special circumstances. Is the North Pole up and the South Pole down? Is someone standing at one of the poles right-side up or upside-down? Kind of depends on your perspective.

When I studied how to think in school, I was taught that the first rule of logic was that a thing cannot both be and not be at the same time and in the same respect. That last note, "in the same respect," says a lot. As soon as you change the frame of reference, you've changed the truthiness of a once immutable fact.

Death seems pretty definite. The body is just a lump. Life is gone. But if you step back a bit, the body is actually in a transitional phase while it slowly turns into compost—capable of living in another way.

This is not to say that nothing is true or that everything is possible—just that it might not be so helpful for things to be known as true for all time, without a disclaimer. At the moment, the way it's presented to us, astrology is highly unlikely to be true. But if it turns out that organic stuff once bounced off Mars and hit Earth with a dose of life, we might have to revise some statements that planets don't influence our lives here on Earth.

I wonder—and this is just a modest proposal—whether scientific truth should be identified in a way acknowledging that it's something we know and understand *for now, and in a certain way.* The public comes to mistrust science when they feel that scientists can't make up their minds. One says red wine is good for you and another says even in small amounts it can be harmful. In turn, some people think science is just another belief system.

Scientists and science writers make a real effort to deal with this all the time. The phrase "current research suggests . . ." warns us that it's not a fact yet. But from time to time the full-blown factualness of something is declared, even though further work could place it within a new frame of reference. And then the public might wonder if the scientists are just arguing for their pet ideas.

Facts, it seems to me, are workable units, useful in a given frame or context. They should be as exact and irrefutable as possible, tested by experiment to the fullest extent. When the frame changes, they don't need to be discarded as untrue but respected as still useful within their domain. Most people who work with facts accept this, but I don't think the public fully gets it.

That's why I hope for more wariness about implying we know something to be true or false for all time and for everywhere in the cosmos.

Especially if we happen to be upside-down when we say it.

SIMPLE ANSWERS

GAVIN SCHMIDT

Climatologist, director, NASA's Goddard Institute for Space Studies

More precisely, the notion that there are simple answers to complex problems. The universe is complicated. Whether you're interested in the functioning of a cell, the ecosystem in Amazonia, the climate of the Earth, or the solar dynamo, almost all the systems and their effects on our lives are complex and multifaceted. It's natural for us to ask simple questions about these systems, and many of our greatest insights have come from the profound examination of such simple questions. However, the answers that have come back are never as simple. The answer in the real world is never "42."

Yet collectively we keep acting as though there were simple answers. We continually read about the search for the one method that will let us cut through the confusion, the one piece of data that tells us the "truth," or the final experiment that will prove the hypothesis. But most scientists agree that these are fool's errands—that science is a method for producing incrementally useful approximations to reality, not a path to absolute truth.

In contrast, our public discourse is dominated by voices equating clarity with seeing things as either good or bad, day or night, black or white. They're not ignoring just the shades of gray; they're missing out on the whole wonderful multi-hued spectrum. By demanding simple answers to complex questions, we rob the questions of the qualities that make them interesting.

Scientists sometimes play into this limiting frame when we craft our press releases or pitch our popular-science books, and in truth it's hard to avoid. But we should be more vigilant. The world is complex, and we need to embrace that complexity to have any hope of finding robust answers to the simple questions we will inevitably continue to ask.

WE'LL NEVER HIT BARRIERS TO SCIENTIFIC UNDERSTANDING

MARTIN REES

Former president, Royal Society; emeritus professor of cosmology and astrophysics, University of Cambridge; master, Trinity College; author, From Here to Infinity

There's a widely held presumption that our insight will deepen indefinitely—that all scientific problems will eventually yield to attack. But we may need to abandon this optimism. The human intellect may hit the buffers—even though in most fields of science, there's surely a long way to go before that happens.

There is plainly unfinished business in cosmology. Einstein's theory treats space and time as smooth and continuous. We know, however, that no material can be chopped into arbitrarily small pieces; eventually you get down to discrete atoms. Likewise, space itself has a grainy and "quantized" structure, but on a scale a trillion trillion times smaller. We lack a unified understanding of the bedrock of the physical world.

Such a theory would bring Big Bangs and multiverses within the remit of rigorous science. But it wouldn't signal the end of discovery. Indeed, it would be irrelevant to the 99 percent of scientists who are neither particle physicists nor cosmologists. Our grasp of diet and child care, for instance, is still so meager that expert advice changes from year to year. This may seem in incongruous contrast to the confidence with which we can discuss galaxies and subatomic particles. But biologists are held up by the problems of complexity—and these are more daunting than those of the very big and the very small.

The sciences are sometimes likened to different levels of a tall building: particle physics on the ground floor, then the rest of physics, then chemistry, and so forth—all the way up to psychology (and the economists in the penthouse). There's a corresponding hierarchy of complexity: atoms, molecules, cells, organisms, and so forth. This metaphor is in some ways helpful; it illustrates how each science is pursued independently of the others. But in one key respect the analogy is poor. In a building, insecure foundations imperil the floors above. But the "higher level" sciences dealing with complex systems aren't imperiled by an insecure base, as a building is.

Each science has its own distinct concepts and explanations. Even if we had a hypercomputer that could solve Schrödinger's equation for quadrillions of atoms, its output wouldn't yield the kind of understanding that most scientists seek.

This is true not only of the sciences that deal with really complex things—especially those things that are alive—but even when the phenomena are more mundane. For instance, mathematicians trying to understand why taps drip, or why waves break, don't care that water is H_2O. They treat the fluid as a continuum. They use "emergent" concepts like viscosity and turbulence.

Nearly all scientists are reductionists, insofar as they think that everything, however complicated, obeys the basic equations of physics. But even if a hypercomputer could solve Schrödinger's equation for the immense aggregate of atoms in (say) breaking waves, migrating birds, or tropical forests, an atomic-level explanation wouldn't yield the enlightenment we really seek. The brain is an assemblage of cells, and a painting is an assemblage of chemical pigment. But in both cases, what's interesting is the pattern and structure—the emergent complexity.

We humans haven't changed much since our remote ancestors roamed the African savannah. Our brains evolved to cope with the human-scale environment. So it's surely remarkable that we can make sense of phenomena that confound everyday intuition, in particular the minuscule atoms we're made of and the vast cosmos that surrounds us. Nonetheless—and here I'm sticking my neck out—maybe some aspects of reality are intrinsically beyond us, in that their comprehension would require some posthuman intellect—just as Euclidean geometry is beyond nonhuman primates.

Some may contest this proposition by pointing out that there's no limit to what's computable. But being computable isn't the same as being conceptually graspable. To give a trivial example, anyone who has learned Cartesian geometry can readily visualize a simple pattern—a line or a circle—when given the equation for it. But nobody given the (simple-seeming) algorithm for drawing the Mandelbrot set could visualize its amazing intricacies, even though drawing the pattern is only a modest task for a computer.

It would be unduly anthropocentric to believe that all of science—and a proper concept of all aspects of reality—is within human mental powers to grasp. Whether the really long-range future lies with organic posthumans or with intelligent machines is a matter for debate—but either way there will be insights into reality left for them to discover.

LIFE EVOLVES VIA A SHARED GENETIC TOOLKIT

SEIRIAN SUMNER

Senior lecturer in behavioral biology, University of Bristol, U.K.

Genes and their interaction networks determine the phenotype of an organism—what it looks like and how it behaves. One of the biggest problems in modern evolutionary biology is understanding the relationship between genes and phenotypes. The prevailing theory is that all animals are built from essentially the same set of regulatory genes—a genetic toolkit—and that phenotypic variation within and between species arises simply by using these shared genes differently. But scientists are now generating a vast amount of genomic data from an eclectic mix of organisms, and these data are telling us to put to bed the idea that all life is underlain by a common toolkit of conserved genes. Instead, we need to examine the role of genomic novelty in the evolution of phenotypic diversity and innovation.

The idea of a conserved genetic toolkit of life comes from the "evo-devo" (evolutionary and developmental biology) world. In short, it proposes that evolution uses the same ingredients in all organisms but tinkers with the recipe. By expressing genes at different times in development and/or in different parts of the body, the same genes can be used in different combinations to allow evolvability, generating phenotypic diversity and innovation. Animals look different not because the molecular machinery is different but because different parts of the machinery are activated to differing degrees at different times, in different places, and in different combinations. The number of combinations is huge, so this is a plausible explanation for the development of complex and diverse phenotypes from even a small number of genes. For example, we humans have a mere 21,000 genes in our genome, yet we're arguably one of the most complex products of evolution.

A textbook example is the supercontroller of development, *Hox* genes—a set of genes that tell bodies where to develop heads, tails, arms, legs, in every major animal group. *Hox* genes are in mice, worms, humans. . . . They're inherited from a common ancestor. Other examples of toolkit genes are those controlling eye development or hair/plumage coloration. Toolkit genes are old, present in all animals, and do pretty much the same thing in all animals. There's no denying that conserved genomic material forms an important part of the molecular building blocks of life.

However, we can now sequence *de novo* the genomes and transcriptomes (the genes expressed at any one time/place) of any organism. We have sequence data for algae, pythons, green sea turtles, puffer fish, pied flycatchers, platypuses, koalas, bonobos, giant pandas, bottle-nosed dolphins, leafcutter ants, monarch butterflies, Pacific oysters, leeches . . . the list grows exponentially. And each new genome brings with it a suite of unique genes. Twenty percent of genes in nematodes are unique. Each lineage of ants contains about 4,000 novel genes, but only sixty-four of these are conserved across all seven ant genomes sequenced so far.

Many of these unique ("novel") genes are proving important in the evolution of biological innovations. Morphological differences between closely related freshwater polyps, *Hydra*, can be attributed to a small group of novel genes. Novel genes are emerging as important in the worker castes of bees, wasps, and ants. Newt-specific genes may play a role in their amazing tissue-regenerative powers. In humans, novel genes are associated with such devastating diseases as leukemia and Alzheimer's.

Life is genomically complex, and this complexity plays a crucial role in the evolving diversity of life. It's easy to see how an innovation can be improved through natural selection: For example, once the first eye evolved, it was subject to strong selection to increase the fitness (survival) of its owner. It's more challenging to explain how novelty originates, especially from a conserved genomic toolkit. Darwinian evolution explains how organisms and their traits evolve, but not how they originated. How did the first eye arise? Or, more specifically, how did that master regulatory gene for eye development in all animals first originate? The capacity for evolving new phenotypic traits—be they morphological, physiological, or behavioral—is critical for survival and adaptation, especially in changing (or new) environments.

A conserved genome can generate novelties through rearrangements (within or between genes), changes in regulation, or genome-duplication events. For example, the vertebrate genome has been replicated in its entirety twice in its evolutionary history, and salmonid fish have undergone a further two whole genome duplications. Duplications reduce selection on the function of one of the gene copies, allowing that copy to mutate and evolve into a new gene while the other copy maintains business as usual. Conserved genomes can also harbor a lot of latent genetic variation—fodder for evolving novelty—which is not exposed to selection. Nonlethal variation can lie dormant in the genome by not being expressed or by being expressed at times when it doesn't have a lethal effect on the phenotype. The molecular machinery regulating expression of genes and proteins depends on minimal information, rules, and tools: Transcription factors recognize sequences of only a few base pairs as binding sites, which gives them enormous potential for plasticity in where they bind. Pleiotropic changes across many conserved genes using different combinations of transcription, translation, and/or post-translation activity are a good source of genomic novelty. For instance, the evolution of beak shapes in Darwin's finches is controlled by pleiotropic changes brought about by changes in the signaling patterns of a conserved gene that controls bone development. The combinatorial power of even a limited genetic toolkit gives it enormous potential to evolve novelty from old machinery.

But the presence of unique genes in all evolutionary lineages studied to date tells us that *de-novo* gene birth, rather than a reordering of old ingredients, is important in phenotypic evolution. The overabundance of noncoding DNA in genomes is less puzzling if they're a melting pot for genomes to exploit and create new genes and gene function—and ultimately phenotypic innovation. The current thinking is that genomes are constantly producing new genes but that only a few become functional.

Our story started simply: All life is a product of gentle evolutionary tinkering of a shared molecular toolkit. The unimaginable time has arrived when we can unpack the molecular building blocks of any creature. And these data are shaking things up. What a surprise? Not really. Perhaps the most important lesson from this is that no theory is completely right—that good theories are those that keep evolving and embracing innovation. Let's evolve theories (keeping the bits that prove correct), not retire them.

FULLY RANDOM MUTATIONS

KEVIN KELLY

Senior maverick, Wired; *author,* Cool Tools: A Catalog of Possibilities

What is commonly called "random mutation" does not in fact occur in a mathematically random pattern. The process of genetic mutation is extremely complex, with multiple pathways, involving more than one system. Current research suggests most spontaneous mutations occur as errors in the repair process for damaged DNA. Neither the damage nor the errors in repair have been shown to be random in where they occur, how they occur, or when they occur. Rather, the idea that mutations are random is simply a widely held assumption by nonspecialists and even many teachers of biology. There's no direct evidence for it.

On the contrary, there's much evidence that genetic mutation varies in patterns. For instance, it's pretty much accepted that mutation rates increase or decrease as stress on the cells increases or decreases. These variable rates of mutation include mutations induced by stress from an organism's predators and competition, as well as mutations brought on by environmental and epigenetic factors. Mutations have also been shown to have a higher chance of occurring near a place in DNA where mutations have already occurred, creating mutation hotspot clusters—a nonrandom pattern.

Although we can't say mutations are random, we *can* say that there's a large chaotic component, just as there is in the throw of loaded dice. But loaded dice shouldn't be confused with randomness, because over the long run—which is the time frame of evolution—the weighted bias will have noticeable consequences. So, to be clear: The evidence shows that chance plays a primary role in mutations, and there would be no natural selection without chance. But it's not random chance. It's loaded chance, with multiple constraints, multipoint biases, numerous clustering effects, and skewed distributions.

So why does the idea of random mutations persist? The assumption of "random mutation" was a philosophical necessity to combat the erroneous earlier idea of inherited acquired traits, or what's commonly called Lamarckian evolution. As a rough first-order approximation, random mutation works pretty well as an intellectual and experimental framework. But the lack of direct evidence for actual random mutations has now reached a stage where the idea needs to be retired.

There are several related reasons why this unsubstantiated idea continues to be repeated without evidence. The first is fear that nonrandom mutations would be misunderstood and twisted by creationists to deny the reality and importance of evolution by natural selection. The second is that if mutations aren't random and have some pattern, then that pattern creates a micro direction in evolution. And since biological evolution is nothing but micro actions accumulating into macro actions, these micro patterns leave open the possibility of macro directions in evolution. That raises all kinds of red flags. If there are evolutionary macro directions, where do they originate? And what are the directions? To date, there's little consensus about evidence for macro directions in evolution, beyond an increase in complexity. But the very notion of evolution with *any* direction is so contrary to current dogma in modern evolution theory that the assumption of randomness persists.

By retiring the notion of fully random mutations, we can

gain some practical advantages. The idea that mutations have a bias can be exploited to more easily engineer genetic processes using those biases. We can better understand the origin of disease mutations and remedy them. And with this new understanding, we can better resolve some of the remaining mysteries of macro evolution. An important part of retiring the idea of random mutations is to realize that the chance element operating in mutations is not "imperfect" randomness but rather contains a bit of order that's generative—a small something that can be used by either us or natural selection. What it's used for, or can be used for, is wide open, but we'll never get there if we cling to the idea that mutations are random.

ONE GENOME PER INDIVIDUAL

ERIC J. TOPOL

Gary and Mary West Chair of Innovative Medicine, professor of genomics, Scripps Research Institute; author, The Creative Destruction of Medicine

We were taught that the fertilized egg divides to ultimately yield a human being—recently estimated to have some 37 trillion cells, each with the same, authentic copy of one's genome. Unfortunately, that simple, seemingly immutable archetype just got mutated.

Although the classical teaching—one genome per individual—began to be questioned decades ago, only recently, through our newfound ability to perform single-cell sequencing and high-resolution-array genomic hybridization, was it unequivocally debunked. For example, a 2012 study reported that brain cells from thirty-seven of fifty-nine women autopsied had the Y-chromosome-specific *DYS14* gene.[g] Many found that hard to accept. But recently researchers at the Salk Institute did single-cell sequencing of post-mortem human brain neurons and found that a striking proportion of the cells (up to 41 percent) had structural DNA variants. This level of so-called mosaicism in the brain was far greater than anticipated and raised the question of whether our single-cell sequencing technology might have some flaws accounting for the observation. That doesn't appear to be the case, however, as too many independent studies have come up with similar findings, whether in the brain or other organs, such as skin, blood, and heart. Last year, a group headed by Richard Lifton and Martina Brueckner at Yale found that a high fraction of children with congenital heart disease carried mutations not present in either parent, perhaps accounting for 10 percent of severe heart-disease birth defects.[h]

These spontaneous *de novo* mutations of cells in the course of one's life are a curve ball for geneticists, who thought that heritability was a generation-passed-down story. More reports of sporadic disease keep popping up, attributable to these *de novo* mutations, such as amyotrophic lateral sclerosis (Lou Gehrig's disease), autism, and schizophrenia. The mutations can occur at many time points along the human life span. A sample of fourteen aborted human embryos in development showed that 70 percent had major structural variations, even though this would not be representative of live births.[i] At the other end of the time continuum, in six people whose deaths were unrelated to cancer, there was extensive mosaicism across all organs assessed, including liver, small intestine, and pancreas.[j]

But we still don't know whether this is merely of academic interest or has important disease-inducing effects. Certainly the mosaicism occurring later in life in "terminally differentiated" cells is known to be important in the development of cancer. And the mosaicism of immune cells, particularly lymphocytes, appears to be part of a healthy, competent immune system. Beyond this, the functional significance of each of us carrying multiple genomes largely remains unclear.

The implications are potentially big. When we use a blood sample to evaluate a person's genome, we have no clue about the potential mosaicism existing throughout the individual's body. A lot more work needs to be done to sort this out, and now that we have the technology to do it we'll undoubtedly better understand our remarkable heterogeneous genomic selves in the years ahead.

NATURE VERSUS NURTURE

TIMO HANNAY

Managing director, Digital Science, Macmillan; co-organizer, SciFoo

Any number of scientific theories ought to bite the dust. That's what happens when you work at the frontiers of human ignorance. But most of them are at worst minor distractions or intellectual detours that barely escape the cloisters of academe. A scientific misconception that truly deserves a bullet in the back of its head would be one that has escaped into the real world to do real damage there. Perhaps the best current example is the notion of nature versus nurture.

It's a beguiling concept: highly intuitive and expressible through an alliterative, almost poetic moniker. Francis Galton, who was the founder of eugenics, a polymath, and the cousin of Charles Darwin, coined the term. Unfortunately, like Galton's other monumentally bad idea, "nature versus nurture" creates a corrosive blend of conceptual falsehood and political potency.

The most elementary error people make in interpreting the effects of genes versus the effects of the environment is to assume that you can truly separate one from the other. Donald Hebb, the brilliant Canadian neuropsychologist, when asked whether nature or nurture contributed more to human personality, reportedly said, "Which contributes more to the area of a rectangle, its length or its width?"

This was a clever reply, but unfortunately only reinforced the highly misleading idea that genetics and environment are orthogonal concepts, like Newtonian space and time. In fact, they're more like Einsteinian spacetime—deeply intertwined and with complex interactions that can give rise to counterintuitive results.

Of course, the experts already know this. They realize, for example, that most children inherit from their parents not only genes but also their environment—hence studies of separated monozygotic twins (who share most of their genes but not their environments). In addition, the idea of the extended phenotype—in which organisms, driven by their genes, act to modify their environments—has been well understood for over thirty years. And the science of epigenetics, though still very much a work in progress, has already demonstrated a wide variety of ways in which a gene's effects can be altered by factors other than its nucleotide sequence and shown that these are determined in large part by the gene's environment (which, of course, consists in part of other genes, both in the same organism and beyond).

Again unfortunately, most of this is lost on the people, such as journalists and politicians, who seek to shape our society. Almost all of them seem to retain a naïve Newtonian view of nature and nurture, and this leads them into all sorts of intellectual fallacies.

A case in point is the brouhaha accompanying the release in October 2013 of a lengthy screed on education policy written by Dominic Cummings, then advisor to the U.K.'s center-right education minister. Among other things, he pointed out (correctly) that academic performance is highly heritable. This led many commentators, especially on the left, to equate his statement with the belief that education doesn't matter. In their Newtonian nature-versus-nurture universes, the heritability of a trait is an immutable law that can leave people—worse still, children—as prisoners of their genes.

This is nonsense. Inheritability is not the inverse of mutability, and to say that the heritability of a trait is high is not to say that the environment has no effect, because heritability scores are themselves affected by the environment. Take the case of height. In the rich world, the heritability of height is something like 80 percent. But this is only because our nutrition is universally quite good. In places where malnutrition or starvation are common, environmental factors predominate and the heritability of height is much lower.

Similarly, a high heritability of academic performance is not necessarily a sign that education matters little. On the contrary, it's at least partly a product of modern universal schooling. Indeed, if every child received an identical education, the heritability of academic performance would rise to 100 percent (because any differences could be explained only by genes). Looked at in this way, a high heritability of academic performance is not a right-wing belief but rather a left-wing aim. But try explaining that to a newspaper columnist on a deadline or a politician with an ax to grind. Ironically, a central thrust of Cummings' paper was to argue that the British education system has produced an inept political elite and commentariat oblivious to such technical subtleties. In criticizing his comments, they have merely proved him right.

Thus the misguided concept of "nature versus nurture" causes apparently intelligent people to confuse egalitarianism for fascism, to misunderstand the consequences of their own policies, and hence to arrive at unfounded beliefs regarding the education of our children. The only form of evolutionary manipulation that makes sense here is a concerted effort to eliminate this outdated and misleading idea from the meme pool.

THE PARTICULARIST USE OF "A" GENE-ENVIRONMENT INTERACTION

ROBERT SAPOLSKY

Neuroscientist, Stanford University; author, Monkeyluv: And Other Essays on Our Lives as Animals

As the year 2013 closed, media pundits, as is the custom at such a time, suggested a variety of words and terms that should be banned. Some of those most often listed were "YOLO," "bromance," "selfie," "mancave," and (of course; please, God, make it so) "twerking." In these cases, it wasn't that the terms were in some way wrong but just that they'd become ubiquitous and irritating.

Similarly, some terms in the science world beg to be retired. That's rarely the case simply because an expression has been ubiquitous and irritating. "Genomic revolution" might be ubiquitous and irritating, but it's useful. So is "For 99 percent of hominid history . . ." when discussing what humans do in a setting less artificial than our modern world. Personally, I hope the latter phrase won't be retired, as I use it ubiquitously and irritatingly, with no plans to stop.

However, various scientific concepts should be retired because they're just plain wrong. An obvious example, more pseudoscience than science, is that evolution is "just" a theory. But what I propose retiring is a phrase that's right in the narrow sense but carries wrong connotations: This is the idea of "a gene-environment interaction."

The notion of the effects of a particular gene and a particular environment interacting was a critical counter to the millennium-old dichotomy of nature versus nurture. Its utility in that realm most often took the following form: "Such-and-such may not be all genetic. Don't forget that there may be a gene-environment interaction," rather than "Such-and-such may not be all environmental. Don't forget that there may be a gene-environmental interaction."

The concept was especially useful when expressed quantitatively, in the face of behavioral geneticists' attempts to attribute percentages of variability in a trait to environment as opposed to genes. It also was the basis of a useful rule-of-thumb phrase for non-scientists: "But only if. . . ." As in "You can often say that Gene *A* causes Effect *X*, although sometimes it is more correct to say that Gene *A* causes Effect *X* but only if it's in Environment *Z*. In that case, you have something called a gene-environment interaction."

What's wrong with any of that? It's an incalculably large improvement over "Nature or nurture?"—especially when a supposed answer to that question has gotten into the hands of policy makers or ideologues.

My problem with the concept is with the particularist use of "a" gene-environment interaction—the notion that there can be just one. At its most benign, this notion implies that there can be cases where there *aren't* gene-environment interactions. Worse, that those cases are in the majority. Worst, that lurking out there is something akin to a Platonic ideal of every gene's action—that any given gene has an idealized effect, that it consistently "does" that, and that circumstances where that doesn't occur are rare and represent either pathological situations or inconsequential specialty acts. Thus, a particular gene may have a Platonically "normal" effect on intelligence, unless, of course, the individual was protein-malnourished as a fetus, had untreated phenylketonuria, or was raised as a wild child by meerkats.

The problem with "a" gene-environment interaction is that there is *no* gene that does something. Rather, a gene only has a particular effect in a particular environment; to say that a gene has a consistent effect in every environment is really only to say that it has a consistent effect in all the environments in which it has been studied to date. This has become ever clearer in studies of the genetics of behavior, as there has been increasing appreciation of the environmental regulation of epigenetics, transcription factors, splicing factors, and so on. And this is most dramatically pertinent to humans, given the extraordinary range of environments—both natural and culturally constructed—in which we live.

Investigating "a gene-environment interaction" is the same as asking what length has to do with the area of a rectangle and being told that in this particular case there's a length/width interaction.

NATURAL SELECTION IS THE ONLY ENGINE OF EVOLUTION

ATHENA VOULOUMANOS

Associate professor of psychology; principal investigator, NYU Infant Cognition and Communication Laboratory

In evolution classes, Lamarckism—the notion promoted by Lamarck that an organism could acquire a trait during its lifetime and pass that trait to its offspring—is usually only briefly discussed and often ridiculed. Darwin's theory of natural selection is presented as the one true mechanism of evolutionary change.

In Lamarck's famous example, giraffes that ate leaves from higher branches could potentially grow longer necks than giraffes that ate from lower branches and could pass their longer necks on to their offspring. The inheritance of acquired characteristics was originally considered a legitimate theory of evolutionary change, with even Darwin proposing his own version of how organisms might inherit acquired characteristics.

Experimental hints of intergenerational transfer of acquired traits came in 1923, when Pavlov reported that while his first generation of white mice needed 300 trials to learn where he hid food, their offspring needed only 100, and their grandchildren only 30. But Pavlov's description didn't make clear whether or not the mice were all housed together (allowing for some communication between mice) or whether or not they had access to other kinds of learning. Still other early studies of potential intergenerational-trait-transfer in plants, insects, and fish also suffered from alternative interpretations or poorly controlled experiments. Lamarckism was dismissed.

But more recent studies, using modern reproduction techniques like in-vitro fertilization and proper controls, can physically isolate generations from each other and rule out any kind of social transmission or learning. For example, mice that were fear-conditioned to an otherwise neutral odor produced baby mice that also feared that odor. Their grand-baby mice feared it, too. Unlike in Pavlov's studies, communication couldn't be the explanation. Because the mice never fraternized and cross-fostering experiments further ruled out social transmission, the newly acquired specific fear had to be encoded in their biological material. (Biochemical analysis showed that the relevant change was likely in the methylation of olfactory reception genes in the sperm of the parents and offspring. Methylation is one example of an epigenetic mechanism.) Natural selection is still the primary shaper of evolutionary change, but the inheritance of acquired traits may play an important role, too.

These findings fit into the relatively new field of study called epigenetics. Epigenetic control of gene expression contributes to the differential development of cells in a single organism (cells that therefore share the same DNA sequence) into, say, heart cells or neurons. But the last decade has shown actual evidence, and possible mechanisms, for how the environment and the organism's behavior in it might cause heritable changes in gene expression—with no change in the DNA sequence passed onto offspring. In recent years, we have seen evidence of epigenetic inheritance across a wide range of morphological, metabolic, and even behavioral traits.

The intergenerational transmission of acquired traits is making a comeback as a potential mechanism of evolution. It also raises the interesting possibility that better diet, exercise, and education, which we thought couldn't affect the next generation (except, with luck, through good example) could actually do just that.

BEHAVIOR = GENES + ENVIRONMENT

STEVEN PINKER

Johnstone Family Professor, Department of Psychology; Harvard University; author, The Sense of Style: The Thinking Person's Guide to Writing in the 21st Century

Would you say that the behavior of your computer or smartphone is determined by an interaction between its inherent design and the way it is influenced by the environment? It's unlikely; such a statement would not be false, but it would be obtuse. Complex adaptive systems have a nonrandom organization, and they have inputs. But speaking of inputs as "shaping" the system's behavior, or pitting its design against its input, would lead to no insight as to how the system works. The human brain is far more complex, and processes its input in more complex ways, than human-made devices, yet many people analyze it in ways that are too simplistic for our far simpler toys. Every term in the equation is suspect.

Behavior: More than half a century after the cognitive revolution, people still ask whether a behavior is genetically or environmentally determined. Yet neither the genes nor the environment can control the muscles directly. The cause of behavior is the brain. While it's sensible to ask how emotions, motives, or learning mechanisms have been influenced by the genes, it makes no sense to ask this of behavior itself.

Genes: Molecular biologists have appropriated the term "gene" to refer to stretches of DNA that code for a protein. Unfortunately, this sense differs from the one used in population genetics, behavioral genetics, and evolutionary theory—namely, any information carrier that's transmissible across generations and has sustained effects on the phenotype. This includes any aspect of DNA that can affect gene expression, and is closer to what is meant by "innate" than genes in the molecular biologists' narrow sense. The confusion between the two leads to innumerable red herrings in discussions of our makeup, such as the banality that the expression of genes (in the sense of protein-coding stretches of DNA) is regulated by signals from the environment. How else could it be? The alternative is that every cell synthesizes every protein all the time! The epigenetics bubble inflated by the science media is based on a similar confusion.

Environment: This term for the inputs to an organism is also misleading. Of all the energy impinging on an organism, only a subset, processed and transformed in complex ways, has an effect on its subsequent information processing. Which information is taken in, how it's transformed, and how it affects the organism (that is, the way the organism learns) all depend on the organism's innate organization. To speak of the environment "determining" or "shaping" behavior is unperspicuous.

Even the technical sense of "environment" used in quantitative behavioral genetics is perversely confusing. Now, there's nothing wrong with partitioning phenotypic variance into components that correlate with genetic variation (heritability) and with variation among families ("shared environment"). The problem comes from the so-called "nonshared" or "unique" environmental influences. This consists of all the variance attributable to neither genetic nor familiar variation. In most studies, it's calculated as 1 – (heritability + shared environment). Practically, you can think of it as the differences between identical twins who grow up in the same home. They share their genes, parents, older and younger siblings, school, peers, and neighborhood. So what could make them different? Under the assumption that behavior is a product of genes plus environment, it must be something in the environment of one that is not in the environment of the other.

But this category really should be called "miscellaneous/ unknown," because it has nothing necessarily to do with any measurable aspect of the environment, such as one sibling getting the top bunk bed and the other the bottom, or a parent unpredictably favoring one child, or one sibling getting chased by a dog, coming down with a virus, or being favored by a teacher. These influences are purely conjectural, and studies looking for them have failed to find them. The alternative is that this component actually consists of the effects of chance—new mutations, quirky prenatal effects, noise in brain development, and events in life with unpredictable effects.

Stochastic effects in development are increasingly being recognized by epidemiologists, frustrated by such recalcitrant phenomena as nonagenarian pack-a-day smokers and identical twins discordant for schizophrenia, homosexuality, and disease outcomes. They're increasingly forced to acknowledge that God plays dice with our traits. Developmental biologists have come to similar conclusions. The bad habit of assuming that anything not classically genetic must be "environmental" has blinkered behavioral geneticists (and those who interpret their findings) into the fool's errand of looking for environmental effects for what may be randomness in developmental processes.

A final confusion in the equation is the seemingly sophisticated add-on of "gene-environment interactions." This is also designed to confuse. Gene-environment interactions do *not* refer to the fact that the environment is necessary for genes to do their thing (which is true of all genes). It refers to a flipflop effect in which genes affect a person one way in one environment but another way in another environment, whereas an alternative gene has a different pattern. For example, if you inherit allele 1, you are vulnerable: a stressor makes you neurotic. If you inherit allele 2, you are resilient: a stressor leaves you normal. With either gene, if you are never stressed, you're normal.

Gene-environment interactions in this technical sense, confusingly, go into the "unique environmental" component, because they're not the same (on average) in siblings growing up in the same family. Just as confusingly, "interactions" in the commonsense sense—namely, that a person with a given genotype is predictably affected by the environment—goes into the "heritability" component, because quantitative genetics measures only correlations. This confound is behind the finding that the heritability of intelligence increases, and the effects of shared environment decrease, over a person's lifetime. One explanation is that genes have effects late in life, but another is that people with a given genotype place themselves in environments that indulge their inborn tastes and talents. The "environment" increasingly depends on the genes, rather than being an exogenous cause of behavior.

INNATENESS

ALISON GOPNIK

Professor of psychology and philosophy, University of California, Berkeley; author, The Philosophical Baby

It's commonplace, in both scientific and popular writing, to talk about innate human traits, "hardwired" behaviors, or "genes" for everything from alcoholism to intelligence. Sometimes these traits are supposed to be general features of human cognition—sometimes they're supposed to be individual features of particular people. The nature/nurture distinction continues to dominate thinking about development. But it's time for innateness to go.

Of course, for a long time people have pointed out that nature and nurture must interact for a particular trait to develop. But several recent scientific developments challenge the idea of innate traits in a deeper way. It isn't just that it's a little of both, some mix of nurture and nature, but that the distinction itself is fundamentally misconceived.

One development is the important new work exploring what are called epigenetic accounts of development and the new empirical evidence for those epigenetic processes. These studies show the many complex ways that gene expression, which is what ultimately leads to traits, is itself governed by the environment.

Take the maternal mice. Michael Meaney of McGill University and his colleagues took two different but genetically identical strains of mice which normally develop different degrees of intelligence and cross-fostered them—the smart mice mothers raised the dumb mice pups. The result was that the dumb mice developed problem-solving abilities similar to those of the smart ones and this was even passed on to the next generation.[k] So were the mice innately dumb or innately smart? The very question doesn't make sense.

Here's a similar human example. There's increasing evidence for an early temperament difference between "orchids" and "dandelions." Children with some genetic and physiological profiles appear to be more influenced by the environment, both for good and bad. For example, a recent study looked at the level of respiratory sinus arrhythmia—basically, the relation between heart rate and breathing—in at-risk poor children. They discovered that children with high RSA who had secure relationships with their parents had fewer behavior problems later than low-RSA children. But the relationship was reversed for high-RSA children who had difficult relationships—they had more problems than the low-RSA children. So were they innately more or less troubled?

The increasingly influential Bayesian models of human learning, models that have come to dominate recent accounts of human cognition, also challenge the idea of innateness, in a different way. At least since Noam Chomsky, there have been debates about whether we have innate knowledge. The Bayesian picture characterizes knowledge in terms of a set of potential hypotheses about the world. We initially believe that some hypotheses are less probable and others are more so. As we collect new evidence, we can rationally update the probability of these hypotheses. We can discard what initially looked likely and eventually accept ideas that started out as longshots.

If this picture is right, there's some sense in which everything we'll ever think is potentially there from the start. But it's also true that everything we think is subject to revision and change with increasing evidence. From this probabilistic perspective, it also isn't at all clear what it would mean to talk about whether knowledge is innate or learned. You might say instead that some hypotheses initially have a low or high probability of being confirmed by further evidence. But the hypotheses and evidence are inextricably intertwined.

The third development is increasing evidence for a new picture of the evolution of human cognition. The old "Swiss Army knife" picture of capital E capital P "Evolutionary Psychology," with the evolution of myriad different constrained "modules," looks increasingly implausible. The more recent, and more biologically plausible, picture is that more general developmental changes are involved. These include an increase in the Bayesian learning abilities just described, increased cultural transmission, wider parental investment, longer developmental trajectories, and greater capacities for counterfactual thinking. All this leads to feedback loops that rapidly transform human behavior. The evolutionary theorist Eva Jablonka has described the evolution of human cognition as more like the evolution of a hand—a multipurpose flexible tool capable of performing unprecedented behaviors and solving unprecedented problems—than like the construction of a Swiss Army knife. In particular, a number of theorists have argued that the difference between the early emergence of "anatomically modern" humans and the much later emergence of "behaviorally modern" humans is due to these feedback loops rather than to some genetic change.

For example, small changes in the capacity for cultural learning and the period of protected childhood in which that learning can take place could initially lead to small changes in behavior. But the "cultural ratchet" effect could lead to the rapid and accelerating transformation of behavior over generations, especially as there was more and more interaction within groups of early humans.

Combining cultural transmission with Bayesian learning means that each generation of children can integrate the cumulative information of earlier generations. Thus they can imagine alternative ways that the social and physical environment might be structured, and they can implement those changes. So each successive generation will grow up shaped by a new social and physical environment, unlike that of the generations who have gone before—and that in turn will lead each new generation to make new discoveries, reshape the environment again, and so on, in an accelerating process of cognitive and behavioral transformation.

All three of these scientific developments suggest that almost everything we do is not just the result of the interaction of nature and nurture; it is both, simultaneously. Nurture is our nature, and learning and culture are our most important and distinctive evolutionary inheritance.

MORAL BLANK-SLATEISM

KILEY HAMLIN

Director, Centre for Infant Cognition, University of British Columbia

There's a persistent belief in our society that morality is acquired slowly and at considerable effort after birth. It's common to view young children as moral "blank slates," beginning life with no moral leanings of any kind. On this viewpoint, children first encounter the moral world in person, via their own experiences and observations. They then actively (or passively, but fewer scholars believe this today) combine such experiences and observations with advances in impulse control, perspective-taking, and complex reasoning, thus becoming more and more "moral" over time.

Moral blank-slateism should be retired. First, although it works well with a picture of infants as "blooming, buzzing confusions" and of toddlers as selfish egoists, developmental psychological research from (at least) the last decade suggests that neither picture is true. For instance, by three months of age infants can already process prosocial and antisocial interactions between unknown third parties, preferring those people who help, versus hinder, someone else to achieve a goal. Indeed, after viewing such interactions, three-month-olds show a reliable tendency to look at the Helper over the Hinderer, and four-and-a-half-month-olds (who can reach) tend to reach for Helpers. Most strikingly, infants' preferences don't seem to reflect simply preferring those who make good things happen (what we might call an "outcome bias"): In the first year, infants prefer those who harm (not help) those who've hindered others; they also prefer those with helpful intentions even if the outcomes they cause are bad.

Rather than selfish egotism, all kinds of prosocial behaviors begin in infancy—helping, sharing, informing, and so forth. Although these behaviors may result from intensive early socialization, research suggests that infants and toddlers are internally rather than externally motivated to be prosocial. Infants help and give without being prompted, and toddlers will choose to help over doing other (really) fun things. These behaviors may result from different emotional states: Toddlers are negatively aroused by seeing others in need, whereas they find helping others (even at a cost to themselves) emotionally rewarding.

Another reason to retire moral blank-slateism: If you believe that morality is born from experience, then you can attribute differences in moral outcomes to differences in experience. This leads to the notion that we can become appropriately moral given the right (and none of the wrong) inputs. Moral failings, then, result from flawed inputs.

Obviously, experience plays a critical role in moral development. Countless studies indicate some causal relationship between experiences relevant to morality (parenting styles, observed violence, abuse) and moral outcomes. But consider Dylan Klebold and Eric Harris, the shooters at Columbine High School in 1999. They were just the first two of what is now a painfully long list of child mass murderers of children in North America. After Columbine, people said that Dylan and Eric played too many violent video games, were bullied in school, or even that their parents hadn't bothered to teach them right from wrong. The first two claims were certainly true (probably not the third), but the rate of video-game-playing and being bullied in childhood is high. What about the 99.9999 percent of children today who do *not* shoot up their schools? What was different about Eric and Dylan?

Eric was a psychopath. Psychopaths are extremely low in empathy and (perhaps) as a result don't mind killing people for fun; the rate of psychopaths in the population of murderers is much higher than average. Psychopathy is a developmental disorder and is considered one of the least treatable of the mental illnesses. Curiously, it's also one of the latest diagnosed—typically not until adolescence or adulthood. Since we know that interventions need to be started early to be effective (think of recent gains in autism treatment from earlier diagnosis), it may be a given that a late-diagnosed disorder would not be susceptible to intervention. My worry, in a nutshell, is that moral blank-slateism's focus on experience makes us reluctant to identify enduring, temperament-based predictors of antisociality in our children. And when we do, it's too late to treat them. It's not that I don't share the reluctance to "pigeonhole" kids—but blaming individual differences on varied experience may be preventing us from using experience to level the playing field through intervention.

Other studies show a link between very early measures of a lack of empathy and antisocial behavior later in life within the typically developing population. These measures usually involve someone acting distressed in front of the infant and determining whether or not the infant looks on with concern/distress. Most infants do, most of the time. A recent study found that non-abused six- to fourteen-month-olds who showed disregard for others' distress were significantly more likely to be antisocial as adolescents.[1] This result suggests that—outside of psychopathy per se—warning signs for antisocial behavior may emerge early, before experience has played much of a role.

Several studies have now documented that experience may influence moral outcomes via a gene-environment interaction. That is, rather than a simple equation in which, say, adverse experiences lead to antisocial children (child + abuse – ameliorating experiences = violence), the relationship between abuse and antisocial behavior is observed only in children with particular versions of various genes known to regulate certain social hormones. So whether they've been abused or not, children with the "safe" gene alleles are all about equally unlikely to engage in antisocial behavior. Children with the "at risk" alleles, on the other hand, are more susceptible to the consequences of abuse.

The common view of infants as moral blank slates has led to a mistaken view of infancy and how moral behavior and cognition work. Understanding how moral development begins, and understanding all of the causes of individual differences, better equips us to address various moral-developmental paths. Moral blank-slateism should be retired.

ASSOCIATIONISM

OLIVER SCOTT CURRY

Departmental lecturer, Institute of Cognitive and Evolutionary Anthropology, University of Oxford

How do birds fly? How do they stay up in the air? Suppose a textbook told you that the answer was "levitation" and proceeded to catalog the different types of levitation (stationary, mobile), its laws ("What goes up must come down," "Lighter things levitate longer") and constraints (quadrupedalism). You'd rapidly realize that flying was not well understood and also that the belief in levitation was obscuring the need for, and holding back, a proper scientific account of aerodynamics.

Unfortunately, a similar situation applies to the question "How do animals learn?" Textbooks will tell you that the answer is "association" and will proceed to catalog the various types (classical, operant), its laws (Rescorla–Wagner), and constraints (autoshaping, differential conditionality, blocking). You'll be told that association is the ability of organisms to make connections between a given stimulus and a given outcome or response—the sound of a bell with the arrival of food, or the left branch of a maze with the administration of pain—merely through (repeated) exposure to their pairing. And you'll be told that because association treats all stimuli equally, it can in principle enable an organism to learn anything.

The problem is that, as with levitation, no one has ever set out a mechanism that could perform such a feat. And no one ever will, because such a mechanism is not possible in theory and hence not possible in practice. At any given time, an organism is confronted by an infinite number of potential stimuli and, subsequently, an infinite number of potential outcomes. A day in the life of a rat, for example, might include waking up, blinking, walking east, twitching its nose, being trampled on, eating a berry, hearing a rumbling noise, sniffing a mate, experiencing a temperature of 41°F, being chased, watching the sun go down, defecating, feeling nauseous, finding its way home, having a fight, going to sleep, and so on. How does the rat discern that of all the possible combinations of stimuli and outcomes, it was the berry alone that made it feel sick? Just as answers presuppose a question, data presuppose a theory. In the absence of a prior theory specifying what to look for and which relationships to test, there's no way of sorting through this chaos to identify useful patterns. And yet what's the defining feature of associative learning? It is the absence of a prior theory. So, like levitation, association is hollow—a misleading re-description of the very phenomenon in need of explanation.

Critics have for centuries pointed out this problem with associationism (sometimes called the problem of induction or the frame problem). And in recent decades there have been countless empirical demonstrations that animals—ants learning their way home, birds learning songs, rats learning to avoid food—don't learn in the way that associationism suggests. But associationism (whether as empiricism, behaviorism, conditioning, connectionism, or plasticity) refuses to die and keeps rising again, albeit encrusted by ever more ad-hoc exceptions, anomalies, and constraints. Its proponents refuse to abandon it, perhaps because they believe there's no alternative.

But there is. In communication theory, information is the reduction of prior uncertainty. Organisms are "uncertain" because they're composed of conditional adaptations that adopt different states under different conditions. These mechanisms can be described in terms of the decision rules they embody—"If A, then B," or "If you detect light, then move toward it." Uncertainty about which state to adopt (to B or not to B) is resolved by attending to the specified conditions (A). The reduction of uncertainty by one-half constitutes one "bit" of information, and so a single decision rule is a one-bit processor. By favoring adaptations with more branching decision rules, natural selection can design more sophisticated organisms, which engage in more sophisticated information processing, asking more questions of the world before coming to a decision. This framework explains how animals acquire information and learn from their environments. For the rat, a rule is, "If you ate something and subsequently felt sick, then avoid that food in future." It has no such rule fingering sunsets, nose twitching, or fighting, which is why it never makes those connections. Similarly, this account explains why organisms facing different ecological problems, composed of different clusters of such mechanisms, are able to learn different things.

So much for rats. What about humans, who obviously can learn things that natural selection never prepared them for? Surely we must be able to levitate? Not at all; the same logic of uncertainty and information processing must apply. If humans can learn novel things, then this must be because they can generate novel uncertainty—to invent, imagine, create new theories, hypotheses, and predictions, and hence to ask new questions of the world. How? The most likely answer is that humans have a range of innate ideas about the world (to do with color, shape, forces, objects, motion, agents, and minds), which they can recombine (almost at random) in an endless variety of ways (as when we dream) and then test these novel conjectures against reality (by means of the senses). And successful conjectures are themselves recombined and revised to build ever more elaborate theoretical systems. So, far from constraining learning, our biology makes it possible: providing the raw materials, guiding the process to a greater or lesser degree, liberating us to think altogether unprecedented thoughts, and fostering the growth of knowledge. This is how we learn from experience—and all without a whiff of association.

Look, nobody disputes that birds fly; the only question is how. Similarly, nobody disputes that humans and other animals learn; the only question is how. Working out the alternative account of learning will involve identifying which innate ideas humans possess, what rules are used to combine them, and how they're revised. But for this to happen, we must first recognize not only that association is not the answer but that association is not even *an* answer. Only then will the science of learning stop levitating and take off for real.

RADICAL BEHAVIORISM

SIMON BARON-COHEN

Professor of developmental psychopathology; director, Autism Research Centre, Cambridge University; author, The Science of Evil

Every student of psychology is taught that radical behaviorism was displaced by the cognitive revolution because it was deeply flawed scientifically. Yet it's still practiced in animal behavior modification and even in some areas of contemporary human clinical psychology. Here I argue that the continued application of radical behaviorism should be retired not just on scientific but also on ethical grounds.

The central idea of radical behaviorism—that all behavior can be explained as the result of learned associations between a stimulus and a response, reinforced or extinguished through reward and/or punishment—stems from the early 20th-century psychologists B. F. Skinner (at Harvard) and John B. Watson (at Johns Hopkins). Radical behaviorism came under public attack when Skinner's book *Verbal Behavior* (1957) received a critical review by cognitivist/linguist Noam Chomsky in 1959 in the journal *Language*. One of Chomsky's scientific arguments was that no amount of exposure to language, and no amount of reward and reinforcement, was going to lead a dog to talk or understand language; whereas for a human infant, despite all the noise in different environments, language learning universally unfolds. This implies that there's more to behavior than just learned associations. There are evolved neurocognitive mechanisms.

At times, this debate was portrayed as if it were between nativism (Chomsky clearly stated that just as an embryo grows, so language unfolds, under a universal genetic program) and empiricist proponents of *tabula rasa* (Skinner was painted as if he believed the newborn human mind was no more than a blank slate, although this was something of a straw man, since in at least one interview Skinner clearly acknowledged the role of genetics).

My scientific reason for arguing that radical behaviorism should be retired is not to revisit the now stale nature-nurture debate (all reasonable scientists recognize that an organism's behavior is the result of an interaction of these) but rather because radical behaviorism is scientifically uninformative. Behavior by definition is the surface level, so it follows that the same piece of behavior could be the result of different underlying cognitive strategies, different underlying neural systems, and even different underlying causal pathways. Two individuals can show the same behavior but can have arrived at it through very different underlying causal routes. Think of a native speaker of English versus someone who has acquired total fluency in English as a second language, or think of a person who is charmingly polite because he is genuinely considerate of others versus a psychopath who has learned how to flawlessly perform being charmingly polite. Identical behavior, produced via different routes. Without reference to underlying cognition, neural activity, and causal mechanisms, behavior is scientifically uninformative.

Given these scientific arguments, you'd have thought radical behaviorism would have been retired long ago, and yet it continues to be the basis of "behavior modification" programs, in which a trainer aims to shape another person's or an animal's behavior, rewarding them for producing surface behavior while ignoring their underlying evolved neurocognitive makeup. Over and above the scientific reasons for retiring radical behaviorism, I have an ethical reason, too.

Lori Marino at Emory University has conducted research at the interface of neuroscience and ethics and examined the life of an orca (a "killer whale") captured in 1983 in Iceland, brought to Sealand of the Pacific, a theme park in British Columbia, and later moved to SeaWorld Orlando in Florida. The orca was trained to do tricks, such as nodding his head in imitation of the trainer nodding her head, or waving his fin in imitation of the trainer waving her hand. The orca dutifully produced the behaviors to get the rewards (food), but over the years in captivity he was involved in the deaths of three people. It has never been documented that orcas have killed a human in the wild, so this may have been a reaction to the radical behaviorists who were training this orca to show new behaviors while ignoring millions of years of evolved social and emotional neurocognitive circuitry in the animal's brain—circuitry that doesn't just vanish in captivity.

Orcas are highly social. They live in family groups and complex societies comprised of "clans," each with its own unique vocalization dialect that likely functions to strengthen group identity. They hunt in groups, a sign of their remarkable capacity for social coordination, and both males and females contribute to child care. Kidnapping one individual orca and placing it in captivity not only isolates the animal from its social community but also reduces its life expectancy and causes signs of ill health, such as the frequent collapse of the dorsal fin. The use of radical behaviorism toward such animals in captivity is doubly unethical, because of the lack of respect for the animal's real nature. The focus on shaping surface behavior ignores who or what the animal really is.

There may be ethical lessons here when we think about the still widespread use of behavior modification of humans in contemporary clinical settings. We need to respect how people think and feel—respect their real nature, rather than simply focusing on whether they can be trained to change their surface behavior.

"INSTINCT" AND "INNATE"

DANIEL L. EVERETT

Linguistic researcher; dean of arts and sciences, Bentley University; author, Language: The Cultural Tool

The idea that human behavior is guided by highly specific innate knowledge has passed its sell-by date. The interesting scientific questions do not encompass either "instinct" or "innate."

This is true for a number of reasons. First, there's never a period in the development of individuals, from their gamete stage to adulthood, when they're not being affected by their environment. It's misguided to think that newborns of any species begin learning from their environment only when they're born. Their cells have been thoroughly bathed in their environment before their parents mated—a bath whose properties are determined by their parents' behavior, environment, and so on. The effects of the environment on development are so numerous, unstudied, and untested in this sense that we currently have no basis for distinguishing environment from innate predispositions or instincts.

Another reason for doubting the usefulness of terms like "insinct" and "innate" is that much of what we believe to be instinctual can change radically when the environment changes radically—even aspects of the environment we might not have thought relevant. For example, in 2004 a group of scientists carried out experiments on rats in the low-gravity environment of Earth orbit. What they discovered was that the self-righting (roughly, the way in which they come to their feet) routine that many had thought to be instinctual was ineffective in low gravity. But the rats didn't simply fail to self-right. They "invented" a new strategy that worked while they were weightless. They showed behavioral flexibility where none had previously been expected.

In any case, the strongest reason for retiring instinct and innateness from scientific thought is the devil of the details, which shows them to be, well, useless. Here's a partial list of possible definitions of "innate" (borrowing from work by King's College London philosopher Matteo Mameli):

(1) A trait is innate if it's not acquired;
(2) A trait is innate if it's present at birth;
(3) A trait is innate if it reliably appears during a particular, well-defined stage of life;
(4) A trait is innate if it's genetically determined;
(5) A trait is innate if it's genetically influenced;
(6) A trait is innate if it's genetically encoded;
(7) A trait is innate if its development doesn't involve the extraction of information from the environment;
(8) A trait is innate if it's not environmentally induced;
(9) A trait is innate if it's not possible to produce an alternative trait by means of environmental manipulations;
(10) A trait is innate if all environmental manipulations capable of producing an alternative trait are abnormal;
(11) A trait is innate if all environmental manipulations capable of producing an alternative trait are statistically abnormal;
(12) A trait is innate if all environmental manipulations capable of producing an alternative trait are evolutionarily abnormal;
(13) A trait is innate if it's highly heritable;

(14) A trait is innate if it's not learned;
(15) A trait is innate if (a) the trait is psychologically primitive and (b) the trait results from normal development;
(16) A trait is innate if it's generatively entrenched in the design of an adaptive feature;
(17) A trait is innate if it's environmentally canalized, in the sense that it's insensitive to some range of environmental variation;
(18) A trait is innate if it's species-typical;
(19) A trait is innate if it's pre-functional. And so on.

All of these definitions have been shown to be inadequate.

But let's suppose we *can* find a workable definition of "instinct" or "innate." We'd still not be ready to use these terms, because we can't attribute something to the human genotype without some evolutionary account of how it might have gotten there. And such an account would have to offer a scenario by which the trait could have been selected. To do that, we'd need information about the extent and character of variation in ancestral forms, as well as the differential survivorship and reproduction of those forms. To know how something was selected, we need to know something about the ecology within which the selection took place, such as what were/are the ecological factors explaining the innate trait either in the biological, social, or other abiotic environment. Next, we'd need to know how the traits could be passed on to subsequent generations. There should be a correlation between phenotypic traits of parents and offspring greater than chance. Then we'd need to know about the population structure during the time of selection. Any evolutionary biologist also knows that we need information concerning population structure, gene flow, and the environment leading to the diffusion of the trait.

We don't know the answers to those questions. We're in no position at present to know the answers. And we'll *never* be able to know some of the answers. Therefore, there simply is no utility to the terms "instinct" and "innate." Let's retire them so the real work can begin.

ALTRUISM

TOR NØRRETRANDERS

Science writer; consultant; lecturer, Copenhagen; author, The Generous Man: How Helping Others Is the Sexiest Thing You Can Do

The concept of altruism is ready for retirement.

Not that the phenomenon of helping others and doing good to other people is about to go away—not at all. On the contrary, the appreciation of the importance of bonds between individuals is on the rise in the modern understanding of animal and human societies. What needs to go away is the basic idea behind the concept of altruism—that there is a conflict of interest between helping yourself and helping others.

The word "altruism" was coined in the 1850s by the great French sociologist Auguste Comte. What it means is that you do something for other people (the Old French *altrui,* from the Latin *alter*), not just for yourself. Thus it opposes egoism or selfishness. This concept is rooted in the notion that human beings (and animals) are dominated by selfishness and egoism, so that you need a concept to explain why they sometimes behave unselfishly and kindly to others.

But the reality is different: Humans are deeply bound to other humans, and most actions are reciprocal and in the interest of both parties (or, in the case of hatred, in the disinterest of both). The starting point is neither selfishness nor altruism but the state of being bound together. It's an illusion to believe that you can be happy when no one else is. Or that other people will not be affected by your unhappiness.

Behavioral science and neurobiology have shown how intimately we're bound. Phenomena like mimicry, emotional contagion, empathy, sympathy, compassion, and prosocial behavior are evident in humans and animals alike. We're influenced by the well-being of others in more ways than we normally care to think of. Therefore a simple rule applies: *Everyone feels better when you're well, and you feel better when everyone is well.*

This correlated state is the real one. Egoism and its opposite concept, altruism, are second-order concepts—shadows or even illusions. This applies also to the immediate psychological level: If helping others fills you with a rewarding "warm glow," as it is called in experimental economics, is it not also in your own interest to help others? Are you not, then, helping yourself? Being kind to others means being kind to yourself.

Likewise, if you feel better and make more money when you're generous and contribute to the well-being and resources of other people—as in the welfare societies that, like my own Denmark, became rich through sharing and equality—then whoever wants to keep everything for himself, with no gift-giving, no taxpaying, and no generosity, is just an amateur egoist. Real egoists share.

It's not altruistic to be an altruist—just wise. Helping others is in your own interest. We don't need a concept to explain that behavior. Auguste Comte's concept is therefore ready for retirement. And we can all just help each other, without wondering why.

THE ALTRUISM HIERARCHY

JAMIL ZAKI

Assistant professor of psychology, Stanford University

Human beings are the unequivocal world champions of niceness. We act kindly not only toward people who belong to our own social groups or can reciprocate our generosity but also toward strangers thousands of miles away who will never know we helped them. All around the world, people sacrifice their resources, well-being, and even their lives in the service of others.

For behavioral scientists, the great and terrible thing about altruism—behavior that helps others at a cost to the helper—is its inherent contradictions. Prosocial behaviors appear to contradict economic and evolutionary axioms about how humans should behave: selfishly, nasty and brutish, red in tooth and claw, or whichever catchphrase you prefer. After all, how could organisms that sacrifice for others survive, and why would nature endow us with such self-defeating tendencies?

In recent decades, researchers have largely solved this problem, offering reasons that perfectly self-oriented organisms might behave altruistically. Solving the "altruism paradox" becomes trivial when individuals help family members (thus advancing helpers' genes) or others who can reciprocate (increasing helpers' chances of future gains) or others in public (enhancing helpers' reputations). We see these motives at work all around us—in parenting, favors for bosses, opera patrons donating just enough to get their names on the donor plaque in the lobby.

More recently, my colleagues and I, as well as other neuroscientists, uncovered another "selfish" motive for altruism:

Helping others simply feels good. It engages brain structures associated with reward and motivation, like those that come online when you see a beautiful face, win money, or eat chocolate. Further, this "reward-related" brain activity not only accompanies giving but also predicts people's willingness to give, suggesting a tight link between pleasure and generosity. This doesn't mean that altruism is the psychological equivalent of Ben and Jerry's, but it does provide converging evidence for UC San Diego economist James Andreoni's idea that generosity produces a hedonic "warm glow."

One common response I receive when presenting this work has grown increasingly bothersome. Often an audience member will claim that if people experience helping as rewarding, then their actions aren't "really" altruistic at all. The claim, as I understand it, traces back to the Kantian notion—embedded in the "cost to the helper" part of altruism's definition—that virtuous action is motivated by principle alone, and that benefiting from that action, whether by material gain or psychological pleasure, disqualifies it as virtuous. Often this contention devolves into long, animated, and (to my mind) useless attempts to find space for true altruism amid an avalanche of ulterior motives.

This altruism hierarchy—with a near-mystical "true" altruism residing somewhere in the distance and our sullied attempts at it crowding out real life—is widespread. It also plays a role in matters of judgment. A recent study by George Newman and Daylian Cain of the Yale School of Management demonstrates that people judge people as less moral when they act altruistically and gain in the process than when they gain from clearly nonaltruistic behavior.[m] In essence, people view "tainted altruism" as worse than no altruism at all.

The altruism hierarchy should be retired. I believe that people often help others absent the goal of any personal gain. The social psychologist Dan Batson, philosopher Philip Kitcher, and others have done the philosophical and empirical work of distinguishing other-oriented and self-oriented motives for prosociality. But I also believe that the reservation of terms such as "pure" or "real" for actions bereft of any personal gain is less than useful—for two reasons, both of which connect to the broader idea of self-negation.

First, the altruism hierarchy is *logically* self-negating. Attempts to identify true altruism often boil down to redacting motivation from behavior altogether. The story goes that in order to be pure, helping others must dissociate from personal desire (to look good, feel rewarded, etc.). But it's logically fallacious to think of *any* human behavior as amotivated. *De facto*, when people engage in actions, it's because they want to. This could represent an overt desire to gain personally, but it could also stem from previous learning (for instance, that helping others in the past has felt good or provided personal gain) that translates into an intuitive prosocial preference. Disqualifying self-motivated behavior as altruistic obscures the universality of motivation in producing all behavior, generous or not.

Second, the altruism hierarchy is *morally* self-negating. Critics of "impure" altruism often seem to chide helpers for acting in human ways—for instance, by doing things that feel good. The ideal, then, seems to entail acting altruistically while not enjoying those actions one bit. To me, this is no ideal at all. It's profound and downright beautiful to think that our core emotional makeup can be tuned toward others, causing us to feel good when we do. Color me selfish, but I'd take that impure altruism over an de-enervated, floating ideal any day.

HUMANS ARE BY NATURE SOCIAL ANIMALS

ADAM WAYTZ

Psychologist; assistant professor of management and organizations, Kellogg School of Management, Northwestern University

For reinforcing a perilous social/psychological imperialism toward other behavioral sciences, and for suggesting that humans are naturally oriented toward others, the strong interpretation of Aristotle's famous aphorism needs to be retired. Certainly sociality is a dominant force that shapes thought, behavior, physiology, and neural activity. However, enthusiasm over the social brain, social hormones, and social cognition must be tempered with evidence that being social is far from easy, automatic, or infinite. This is because our (social) brains, (social) hormones, and (social) cognition on which social processes rely must first be triggered before they do anything for us.

One of the most compelling pieces of evidence for humans' ostensibly automatic social nature comes from Fritz Heider and Mary-Ann Simmel's famous 1944 animation of two triangles and a circle orbiting a rectangle.[n] The animation depicts merely shapes, yet people find it nearly impossible not to construe these objects as human actors and to construct a social drama around their movements. A closer look at the video, and a closer reading of Heider and Simmel's article describing the phenomenon, suggests that the perception of these shapes in social terms is not automatic but must be evoked by features of the stimuli and situation. These shapes were designed to move in trajectories that specifically mimic social behavior. If the shapes' motion is altered or reversed, they fail to elicit the same degree of social responses. Furthermore, participants in the original studies of this animation were prompted to describe the shapes in social terms by the language and instructions the experimenters used. Humans may be ready and willing to view the world through a social lens, but they don't do so automatically.

Despite possessing capacities far beyond those of other animals to consider others' minds, to empathize with others' needs, and to transform empathy into care and generosity, we fail to employ these abilities readily, easily, or equally. We engage in acts of loyalty, moral concern, and cooperation primarily toward our inner circles, at the expense of people outside those circles. Our altruism is not unbounded; it is parochial. In support of this phenomenon, the hormone oxytocin, long considered to play a key role in forming social bonds, has been shown to facilitate affiliation toward one's in-group but can increase defensive aggression toward one's out-group. Other research suggests that this self-sacrificial intragroup love coevolved with intergroup war, and that societies who most value loyalty tend to be those most likely to endorse violence toward out-groups.

Even our arguably most important social capacity, theory of mind—the ability to adopt others' perspectives—can increase competition as much as it increases cooperation, highlighting the emotions and desires of those we like but also highlighting the selfish and unethical motives of those we dislike. For us to consider the minds of others in the first place requires motivation and the necessary cognitive resources. Because motivation and cognition are finite, so too is our social capacity. Thus, any intervention that intends to increase consideration of others—in terms of empathy, benevolence, and compassion—is limited in its ability to do so. At some point, the well of working memory on which our most valuable social abilities rely will run dry.

Because our social capacities are largely nonautomatic, in-group-focused, and finite, we can retire the strong version of Aristotle's statement. At the same time, the concept of humans as "social by nature" has lent credibility to numerous significant ideas: that humans need other humans to survive, that humans tend to be perpetually ready for social interaction, and that studying the specifically social features of human functioning is profoundly important.

EVIDENCE-BASED MEDICINE

GARY KLEIN

Psychologist; senior scientist, MacroCognition LLC; author, Seeing What Others Don't

Any enterprise has its limits and boundary conditions, and science is no exception. When the reach of science moves beyond these boundary conditions, when it demands respect and obedience it hasn't earned, the results can be counterproductive. One example is evidence-based medicine (EBM), which is the scientific idea I think we should retire.

The concept behind EBM is certainly admirable: a set of best practices validated by rigorous experiments. EBM seeks to provide health care practitioners with treatments they can trust—treatments that have been evaluated by randomized controlled trials, preferably blinded. EBM seeks to transform medicine into a scientific discipline rather than an art form. What's not to like? We don't want to return to the days of quack fads and unverified anecdotes.

But we should trust EBM only if the science behind best practices is infallible and comprehensive, and that's certainly not the case. Medical science is not infallible. Practitioners shouldn't believe a published study just because it meets the criteria of randomized-controlled-trial design. Too many of these studies cannot be replicated. Sometimes the researcher got lucky and the experiments that failed to replicate the finding never got published or even submitted to a journal (the so-called publication bias). In rare cases, the researcher has faked the results. Even when the results can be replicated, they shouldn't automatically be believed; conditions may have been set up in a way that misses the phenomenon of interest, so a negative finding doesn't necessarily rule out an effect.

And medical science is not comprehensive. Best practices often take the form of simple rules to follow, but practitioners work in complex situations. EBM relies on controlled studies that vary one thing at a time, rarely more than two or three. Many patients suffer from multiple medical problems, such as type-2 diabetes compounded with asthma. The protocol that works with one problem may be inappropriate for the others. EBM formulates best practices for general populations, but practitioners treat individuals and need to take individual differences into account. A treatment that's generally ineffective might still be useful for a subset of patients. Further, physicians aren't finished once they select a treatment; they often have to adapt it. They need expertise to judge whether a patient is recovering at an appropriate rate. Physicians have to monitor the effectiveness of a treatment plan and then modify or replace it if it isn't working well. A patient's condition may naturally fluctuate, and physicians have to judge the treatment effects on top of this noisy baseline.

Sure, scientific investigations have done us all a great service by weeding out ineffective remedies. For example, a recent placebo-controlled study found that arthroscopic surgery provided no greater benefit than sham surgery for patients with osteoarthritic knees. But we're also grateful for all the surgical advances of the past few decades (hip and knee replacements, cataract treatments) that were achieved without randomized controlled trials and placebo conditions. Controlled experiments are therefore unnecessary for progress in new types of treatments, and they're not sufficient for implementing treatments with individual patients, who each have unique profiles.

Worse, reliance on EBM can impede scientific progress. If hospitals and insurance companies mandate EBM, backed up by the threat of lawsuits when adverse outcomes are accompanied by any departure from best practices, physicians will become reluctant to try alternative treatment strategies that haven't yet been evaluated using randomized controlled trials. Scientific advancement can be stifled if front-line physicians who blend medical expertise with respect for research are prevented from exploration and discouraged from making discoveries.

LARGE RANDOMIZED CONTROLLED TRIALS

DEAN ORNISH

Founder and president, Preventive Medicine Research Institute; clinical professor of medicine, University of California, San Francisco

It's a commonly held but erroneous belief that a larger study is always more definitive than a smaller one, and a randomized controlled trial is the gold standard. However, there's a growing awareness that size doesn't always matter and that a randomized controlled trial may introduce its own biases. We need more creative experimental designs.

In any scientific study, the question is, "What is the likelihood that observed differences between the experimental group and the control group are due to the intervention or due to chance?" By convention, if the probability is less than 5 percent that the results are due to chance, then the finding is considered statistically significant—i.e., a real finding.

A randomized controlled trial, or RCT, is based on the idea that if you randomly assign subjects to an experimental group that receives an intervention or a control group that does not, then any differences (known or unknown) between subjects that might bias the study are as likely to affect one group as the other. While that sounds good in theory, in practice an RCT can often introduce its own set of biases and thus undermine the validity of the findings.

For example, an RCT may be designed to determine whether dietary changes may prevent heart disease and cancer. Investigators identify patients who meet certain selection criteria—for example, they have several risk factors for heart disease or cancer. The study is described to prospective participants in great detail, and they are asked, "If you are randomly assigned to the experimental group, would you be willing to change your lifestyle?" In order to be eligible for the study, the patient needs to answer, "Yes."

However, if that patient is subsequently randomly assigned to the control group, she may begin to make lifestyle changes on her own, since she has already been told in detail what those lifestyle changes are. If the study is of a new drug available only to the experimental group, then this is less of an issue. But in a study of behavioral interventions, patients randomly assigned to the control group are likely to make at least some of those behavioral changes because they believe the changes are worth making or they wouldn't be of interest to the investigators. Or they may be disappointed at having been assigned to the control group and so more likely to drop out of the study, creating selection bias.

In a large-scale RCT, moreover, it's often hard to provide all the subjects in the experimental group with enough support and resources to make the prescribed lifestyle changes. Thus adherence to these changes is often less rigorous than investigators predicted based on earlier pilot studies with smaller groups of patients.

The net effect of all this is to (a) reduce the likelihood that the experimental group will make the desired lifestyle changes and (b) increase the likelihood that the control group will make similar lifestyle changes. This narrows the differences between the two groups and produces a less statistically significant outcome. Thus the conclusion that the behavioral changes had little or no effect may be misleading. This is known as a type-2 error, meaning that there was a real difference but that these design issues obscured its detection.

That's just what happened in the Women's Health Initiative study of dietary modification, which followed nearly 50,000 middle-aged and older women for more than eight years. The women in the experimental group were asked to eat less fat and more fruits, vegetables, and whole grains daily, to see if this could help prevent heart disease and cancer. The women in the control group were not asked to change their diets. However, the experimental group participants failed to reduce their dietary fat as recommended—over 29 percent of their diet was comprised of fat, not the study's goal of less than 20 percent. Also, they didn't substantially increase their consumption of fruits and vegetables. In contrast, the control group reduced its consumption of fat almost as much and increased its consumption of fruits and vegetables, diluting the between-group differences to the point where they weren't statistically significant. The investigators reported that these dietary changes did not protect against heart disease or cancer, but the hypothesis was not really tested.

Paradoxically, a small study may be more likely to show significant differences between groups than a large one. The Women's Health Initiative study cost almost a billion dollars yet did not adequately test the hypotheses. A smaller study provides more resources per patient, at lower cost, to enhance adherence.

Also, the idea in RCTs that you're changing only one independent variable (the intervention) and measuring one dependent variable (the result) is often a myth. Let's say you're investigating the effects of exercise on preventing cancer. You devise a study wherein you randomly assign one group to exercise and the other group to no exercise. On paper, it appears that you're working with just one independent variable. In practice, however, when you place people on an exercise program you're not just getting them to exercise, you're also affecting other factors that may confound the interpretation of your results. For example, people often exercise with other people, and there's increasing evidence that enhanced social support significantly reduces the risk of most chronic diseases. You're also endowing your study participants with a sense of meaning and purpose, and this also produces therapeutic benefits. And when people exercise, they often begin to eat healthier foods.

We need new, more thoughtful experimental designs and systems approaches that take into account these issues. And new genomic insights will help us to better understand individual responses to treatment, instead of hoping that this variability will be averaged out by the random assignment of patients.

MULTIPLE REGRESSION AS A MEANS OF DISCOVERING CAUSALITY

RICHARD NISBETT

Professor of psychology, University of Michigan; author, Intelligence and How to Get It

Did you know that consuming large amounts of olive oil can reduce your mortality risk by 41 percent? Did you know that if you have cataracts and get them operated on, your mortality risk is lowered by 40 percent over the next fifteen years compared to people with cataracts who don't get them operated on? Did you know that deafness causes dementia?

Those claims and scores like them appear every day in the media. They're usually based on studies employing multiple-regression analyis (MRA). In MRA, a number of independent variables are correlated simultaneously with some dependent variable. The goal is typically to show that variable *A* influences variable *B* "net of" the effects of all the other variables. To put that a little differently, the goal is to show that at every level of variables *C*, *D*, and *E*, an association between *A* and *B* is found. For example, drinking wine is correlated with low incidence of cardiovascular disease, controlling for (net of) the contributions to cardiovascular disease of social class, excess weight, age, etc., etc. Epidemiologists, medical researchers, sociologists, psychologists, and economists are particularly likely to use this technique, although it can be used in almost any scientific field.

The claims—always at least implicit, often explicit—that

MRA can reveal causality are mistaken. We know that the target independent variable (consumption of olive oil, for example) brings along its correlations with many other variables, measured in some inevitably imperfect way or not at all. And the level on each of these variables is "self-selected." Any one of them could be driving the effects on the dependent variable.

Would you think the number of children in a classroom matters for how well schoolchildren learn? It seems reasonable that it would. But a number of MRA studies tell us that—net of average income of families in the school district, size of the school, IQ test performance, city size, geographic location, etc.—average class size is uncorrelated with student performance. The implication: We now know we needn't waste money on decreasing the size of classes.

But researchers have assigned kindergartners through third-graders, by the flip of a coin, to either small classes (thirteen to seventeen per class) or larger classes (twenty-two to twenty-five). The smaller classes showed more improvement in standardized test performance; the effect on minority children was greater than the effect on white children. This is not merely another study on the effects of class size; it *replaces* all the multiple-regression studies on class size.

This is the case because the experimenter selects the level on the target independent variable. This means that the experimental classrooms have equally good teachers on average, equally able students, equal social class of students, etc. Thus the only thing that differs between experimental and control classrooms is the independent variable of interest—namely, class size.

MRA studies that attempt to "control" for other factors, such as social class, age, prior state of health, and so on, can't get around the self-selection problem. The sorts of people who get treatment differ from those who don't get it in goodness knows how many ways.

Consider social class. If an investigator wishes to see whether social class is associated with some outcome, anything correlated with social class might be producing or suppressing the effects of class per se. We can be fairly sure that the people consuming all that olive oil are richer, better educated, more knowledgeable about health, and more concerned about health (with spouses also more concerned about their health, and so on). They're less likely to smoke, or to drink to excess, and they probably live in less toxic environments than people who use corn oil. They're also more likely to be of Italian descent (Italians are relatively long-lived) than African descent (blacks have generally high mortality rates). All these variables are candidates for being the true cause of the association between social class and mortality, rather than the consumption of olive oil per se.

Even when there's an attempt to control for all possible variables, they're not necessarily well measured, which means that their contribution to the target dependent variable will be underestimated. For example, there's no unique correct way to measure social class. Education level, income, wealth, and occupational level are all pieces of the pie, and there's no canonical way to weight them to come up with the same social-class value that God has in mind.

A *New York Times* op-ed writer, a PhD at Harvard, recently expressed the opinion that MRA studies are superior to experiments because MRA studies based on Big Data can have many more subjects. The error here is the assumption that having a relatively small number of subjects is likely to mislead.

Larger *n* is always better than smaller *n,* because we're more likely to detect even small effects. But our confidence in studies is based not on the *number* of cases but on whether we have unbiased estimates of effects and the effects are statistically significant. In fact, if you have a statistically significant effect with a relatively small number of subjects, this means, other things equal, that your effect is *bigger* than if it had required a larger number of subjects to reach the same level of significance.

Big Data will be useful for all kinds of purposes, including generating MRA findings that suggest randomized-design experiments that can provide definitive evidence about whether an apparent effect is real. A lovely example of this kind of sequence results from the 2011 finding in MRA research by Guglielmo Becutti and Silvana Pannaina that low levels of sleep are associated with obesity.[o] That finding taken by itself is next to meaningless. Bad health outcomes are almost all correlated with each other: Overweight people have worse cardiovascular health, worse psychological health, use more drugs, get less exercise, and so on. But following the MRA research, experimenters have done the requisite experiments. They deprived people of sleep and found that they did in fact gain weight. Not only that, but researchers found hormonal and endocrine consequences of sleep disturbances that mediated the weight gain.

Multiple regression, like all statistical techniques based on correlation, has a severe limitation: Correlation doesn't prove causation. And no amount of measuring of "control" variables can untangle the web of causality. What nature hath joined together, multiple regression cannot put asunder.

MOUSE MODELS

AZRA RAZA

Professor of medicine; director, MDS Center, Columbia University

An obvious truth, which is either being ignored or going unaddressed in cancer research, is that mouse models don't mimic human disease well and are essentially worthless for drug development. We cured acute leukemia in mice in 1977 with drugs we're still using in exactly the same dosage and duration in humans—with dreadful results. Imagine the artificiality of taking human tumor cells, growing them in lab dishes, then transferring them to mice whose immune systems have been compromised so they can't reject the implanted tumors, and then exposing these "xenografts" to drugs whose killing efficiency and toxicity profiles will then be applied to treat human cancers.

The inherent pitfalls of such a synthesized, unnatural model system have also plagued other disciplines. A recent scientific paper showed that nearly 150 drugs tested, at the cost of billions of dollars, in human trials of sepsis failed because the drugs had been developed using mice. Unfortunately, what looked like sepsis in mice turned out to be very different from what sepsis is in humans. Coverage of this study by Gina Kolata in the *New York Times*[p] brought heated response from within the biomedical research community: "There is no basis for leveraging a niche piece of research to imply that mice are useless models for all human diseases. . . . The key is to construct the appropriate mouse models and design the experimental conditions that mirror the human situation."[q]

The problem is that there *are* no appropriate mouse models that can "mirror the human situation." So why is the cancer research community still dominated by the dysfunctional tradition of employing mouse models to test hypotheses for the development of new drugs?

Robert Weinberg of the Whitehead Institute at MIT has provided the best answer. In an interview, he offered two reasons. First, there's no model with which to replace it, and second, the Food and Drug Administration "has created inertia because it continues to recognize these [models] as the gold standard for predicting the utility of drugs."[r]

There's a third reason, related to the frailties of human nature. Too many eminent laboratories and illustrious researchers have devoted too much of their time to studying malignant diseases in mouse models, and they're the ones reviewing one another's grants and deciding where the National Institutes of Health money gets spent. They're not prepared to concede that mouse models are basically valueless for most cancer therapeutics.

One of the main reasons we stick to this archaic ethos is to obtain funding. Here's one example: I decided to study a bone-marrow malignant disease called myelodysplastic syndromes (MDS), which frequently evolves to acute leukemia, back in the early 1980s. One early decision I made was to concentrate my research on freshly obtained human cells and not rely on mice or petri dishes alone. In the last three decades, I have collected over 50,000 bone-marrow biopsies, normal control buccal smear cells, and blood, serum, and plasma samples, in a well-annotated tissue repository backed by a computerized bank of clinical, pathologic, and morphologic data. Using these samples, we have identified novel genes involved in causing certain types of MDS, as well as sets of genes related to survival, natural history of the disease, and response to therapy. But when I used bone-marrow cells from treated MDS patients to develop a genomic-expression profile that was startlingly predictive of response and applied for an NIH grant to validate the signature, the main criticism was that before confirming it through a prospective trial in humans, I should first reproduce it in mice.

It's time to let go of the mouse models—at least, as surrogates for bringing drugs to the bedside. Remember what Mark Twain said: "What gets us into trouble is not what we don't know, it's what we know for sure that just ain't so."

THE SOMATIC MUTATION THEORY OF CANCER

PAUL DAVIES

Theoretical physicist, cosmologist, astrobiologist; director, BEYOND: Center for Fundamental Concepts in Science, Arizona State University; author, The Eerie Silence: Renewing Our Search for Alien Intelligence

Cancer is one of the most intensively studied phenomena in biology, yet mortality rates from the disease are little changed in decades. Perhaps that's because we're thinking about the problem in the wrong way.

A major impediment to progress is the deep entrenchment of a fifty-year-old paradigm, the so-called somatic mutation theory. It goes like this. A somatic cell serially accumulates genetic damage, eventually reaching a point at which it decouples from the organism's regulatory systems and embarks on its own agenda.

Cancer cells acquire a range of distinctive hallmarks—unfettered proliferation, evasion of apoptosis (programmed cell death), motility and migratory powers, genomic rearrangements, epigenetic alterations, and changes in the mode of metabolism, chromatin architecture, and elasticity (to mention a few)—that collectively confer remarkable robustness and survivability. In the standard picture, cancer, with all these attendant hallmarks, is considered to be reinvented *de novo* in each host organism, the result of a dream run of "lucky" genetic accidents. The gain of all these amazing fitness functions, co-located in the same neoplasm (population of new cells), over a period of as little as months or even weeks, is attributed to a sort of ultra-fast-paced Darwinian evolution going on in the body of the host organism. Unfortunately, this theory, despite its simplicity and popular appeal, has only one successful prediction: that the administration of chemotherapeutic drugs is likely to fail, on account of the neoplasm's ability to rapidly evolve a resistant sub-population.

Armed with the somatic mutation paradigm, the research community has become fixated on the promise of sequencing technology, which enables genetic and epigenetic changes in cells to be measured on a vast scale. If cancer is caused by mutations, so the reasoning goes, then maybe subtle patterns can be teased out of petabytes of bewildering cancer sequencing data. If so, then the answer to cancer—perhaps even that elusive general-purpose cure—might be found by identifying common defects amid all that stunningly complex malfunctioning genetic machinery. Never has science offered a clearer example of a preoccupation with trees at the expense of the forest.

Stand back and take a hard, skeptical look at that forest. Cancer is widespread among multicellular organisms, afflicting mammals, birds, fish, and reptiles. It clearly has deep evolutionary roots, probably stretching back over a billion years to the dawn of multicellularity. Indeed, it represents a breakdown of multicelled cooperation. Unchecked, cancer follows a predictable pattern of progression, usually spreading around the body and colonizing remote organs. It seems to be executing an efficient pre-loaded genetic and epigenetic program. Like a genie in a glass bottle, once it gets out it has a well-defined agenda. Many things can shatter the bottle, but the real culprit is the genie. The cancer research community, unfortunately, is preoccupied with seeking mostly irrelevant patterns amid the random shards of glass, while ignoring the genie.

Why are our cells harboring such dangerous genies? The answer has been known for a long time but is mostly shrugged aside. The same genes active in cancer are also active in early embryogenesis (even in gametogenesis), and to some extent in wound healing and tissue regeneration. These ancient genes are deeply embedded and well protected in our genomes. They run the core functionality of cells. Top of the functionality list is the ability to proliferate—the most fundamental modality of living organisms, with nearly 4 billion years of evolutionary refinement behind it. Cancer seems to be the default state of cells that are stressed or insulted in some way, such as by aging tissue architecture or carcinogenic chemicals, with tumors representing a reversion to an ancestral phenotype.

In biology, few things are black or white. The somatic mutation paradigm is undeniably of some relevance to cancer, and sequencing data is certainly not useless. Indeed, it could prove a gold mine, if only the research community comes to interpret that data in the right way. But the narrow focus of current cancer research is a serious obstacle to progress. Cancer will be understood properly only by positioning it within the great sweep of evolutionary history.

THE LINEAR NO-THRESHOLD (LNT) RADIATION DOSE HYPOTHESIS

STEWART BRAND

Founder, Whole Earth Catalog*; cofounder, The Well, The Long Now Foundation; author,* Whole Earth Discipline

In his 1976 book, *A Scientist at the White House*, George Kistiakowsky, President Eisenhower's science advisor, told what he wrote in his diary in 1960 on being exposed to the idea by the Federal Radiation Council:

> It is a rather appalling document which takes 140 pages to state the simple fact that since we know virtually nothing about the dangers of low-intensity radiation, we might as well agree that the average population dose from man-made radiation should be no greater than that which the population already receives from natural causes; and that any individual in that population shouldn't be exposed to more than three times that amount, the latter figure being, of course, totally arbitrary.

Later in the book, Kistiakowsky, who was a nuclear expert and veteran of the Manhattan Project, wrote: "A linear relation between dose and effect . . . I still believe is entirely unnecessary for the definition of the current radiation guidelines, since they are pulled out of thin air without any knowledge on which to base them."

Sixty-three years of research on radiation effects have

gone by, and Kistiakowsky's critique still holds. The linear no-threshold (LNT) radiation dose hypothesis, which surreally influences every regulation and public fear about nuclear power, is based on no knowledge whatever.

At stake are the hundreds of billions spent on meaningless levels of "safety" around nuclear power plants and waste storage, the projected costs of next-generation nuclear plant designs to reduce greenhouse gases worldwide, and the extremely harmful episodes of public panic that accompany rare radiation-release events like Fukushima and Chernobyl. (No birth defects were caused by Chernobyl, but fear of them led to 100,000 panic abortions in the Soviet Union and Europe.) What people remember about Fukushima is that nuclear opponents predicted that hundreds or thousands would die or become ill from the radiation. In fact nobody died, nobody became ill, and nobody is expected to.

The "linear" part of the LNT is true and well documented. Based on long-term studies of nuclear-industry workers and survivors of the atomic bombing of Japan, the incidence of eventual cancer increases with increasing exposure to radiation at levels above 100 millisieverts/year. The effect is linear. Below 100 millisieverts/year, however, no increased cancer incidence has been detected, either because it doesn't exist or because the numbers are so low that any signal gets lost in the epidemiological noise.

We all die. Nearly half of us die of cancer (38 percent of females, 45 percent of males). If the "no-threshold" part of the LNT is taken seriously and an exposed population experiences as much as a 0.5-percent increase in cancer risk, it simply cannot be detected. The LNT operates on the unprovable assumption that the cancer deaths exist even if the increase is too small to detect, and that therefore "no level of radiation is safe" and every extra millisievert is a public health hazard.

Some evidence against the "no-threshold" hypothesis draws on studies of background radiation. In the U.S., we're all exposed to 6.2 millisieverts a year on average, but it varies regionally. New England has lower background radiation, Colorado is much higher, yet cancer rates in New England are higher than in Colorado—an inverse effect. Some places in the world, such as Ramsar, Iran, have a tenfold higher background radiation, but no higher cancer rates have been discovered there. These results suggest that there is indeed a threshold below which radiation is not harmful.

Furthermore, recent research at the cell level shows a number of mechanisms for repair of damaged DNA and ejection of damaged cells up to significant radiation levels. This isn't surprising, given that life evolved amid high radiation and other threats to DNA. The DNA repair mechanisms that have existed in yeast for 800 million years are also present in humans.

The actual threat of low-dose radiation to humans is so low that the LNT hypothesis can neither be proved true nor proved false, yet it continues to dominate and misguide policies concerning radiation exposure, making them grotesquely conservative and expensive. Once the LNT is explicitly discarded, we can move on to regulations that reflect only discernible, measurable medical effects and respond mainly to the much larger considerations of whole-system benefits and harms.

The most crucial decisions about nuclear power are at the category level of world urban prosperity and climate change, not imaginary cancers per millisievert.

UNIVERSAL GRAMMAR

BENJAMIN K. BERGEN

Associate professor of cognitive science, University of California, San Diego; author, Louder Than Words

The world's languages differ to the point of inscrutability. Knowing the English word "duck" doesn't help you guess the French *canard* or Japanese *ahiru*. But there are commonalities hidden beneath the superficial differences. For instance, human languages tend to have parts of speech (like nouns and verbs). They tend to have ways to embed propositions in other ones. ("John knows that Mary thinks that Paul embeds propositions in other ones.") And so on. But why?

An influential and appealing explanation is known as universal grammar: Core commonalities across languages exist because they're part of our genetic endowment. On this view, humans are born with a predisposition to develop languages with specific properties. Infants expect to learn a language that has nouns and verbs, that has sentences with embedded propositions, and so on. This idea could explain not only why languages are similar but also what it is to be uniquely human—and, indeed, how children acquire their native language. It may also seem intuitively plausible, especially to people who speak several languages: If English (and Spanish and French) have nouns and verbs, why wouldn't every language? To date, universal grammar remains one of the most visible products of the field of linguistics—the one minimally counterintuitive bit that former students often retain from an introductory linguistics class.

But evidence has not been kind to universal grammar. Over the years, field linguists (they're like field biologists with really good microphones) have reported that languages are much more diverse than originally thought. Not all languages have nouns and verbs. Nor do all languages let you embed propositions in others. And so it has gone for basically every proposed universal linguistic feature. The empirical foundation has crumbled out from under universal grammar. We thought there might be universals that all languages share, and we sought to explain them on the basis of innate biases. But as the purportedly universal features have revealed themselves to be nothing of the sort, the need to explain them in categorical terms has evaporated. As a result, what can plausibly make up the content of universal grammar has become progressively modest over time. At present, there's evidence that nothing but perhaps the most general computational principles are part of our innate language-specific human endowment.

So it's time to retire universal grammar. It had a good run, but there's nothing much it can bring us now in terms of what we want to know about human language. It can't reveal much about how language develops in children—how they learn to articulate sounds, to infer the meanings of words, to put words together into sentences, to infer emotions and mental states from what people say, and so on. And the same is true for questions about how humans have evolved or how we differ from other animals. There are ways in which humans are unique in the animal kingdom, and a science of language ought to be trying to understand these. But, again, universal grammar, gutted by evidence as it has been, will not help much.

Of course, it remains important and interesting to ask what commonalities, superficial and substantial, tie together the world's languages. There may be hints there about how human language evolved and how it develops. But to ignore its diversity is to set aside its most informative dimension.

A SCIENCE OF LANGUAGE SHOULD DEAL ONLY WITH "COMPETENCE"

N. J. ENFIELD

Senior staff scientist, Language and Cognition Group, Max Planck Institute for Psycholinguistics, Nijmegen, the Netherlands; author, Relationship Thinking

Suppose that a scientist wants to study a striking animal behavior—say, the courtship display of the stickleback fish or the cooperative agriculture of leafcutter ants. She will, of course, ultimately want to know the underlying mechanisms of these behaviors: How do they work? How did they evolve? What can we learn from them? But no student of animal behavior would dream of asking those questions without first systematically discovering the facts, beginning with extensive field observation in the wild, then moving to experiments and modeling in the lab. Why, then, have linguists emphatically denied any value in directly observing linguistic behavior?

The culprit is a bad idea: that a science of language should be concerned only with competence (the mental capacity for producing sentences) and never with performance (what happens when we actually talk). Here is its decidedly dualist reasoning: When the idealized language patterns tucked away in the mind get "externalized" in communication, they're filtered and shaped by contingencies, such as motor constraints, attention and memory limitations, errors of execution, local conventions, and more. As a result, it is argued, performance bears little useful relation to the predefined object of study: competence. Students of linguistics have been taught not to waste their time with the worldly facts of performance.

This idea amounts to an unaccountably narrow view of what language is. It has diverted linguists' attention from many substantial questions, each with deep implications. Just a few examples: Without looking at performance, we wouldn't see the systematic and ingenious ways in which people handle the constant speech errors, hesitations, and misfires of conversation, along with the social delicacies of navigating these bouts of turbulence. Nor would we see the emerging breakthroughs from statistical research on newly available large language corpora, with results suggesting that we *can* infer competence by looking at performance. Nor, finally, would linguistics have a causal account of how languages evolve historically. In the cycle of language transmission going from public (someone speaks) to private (someone's mental state is affected) and back to public (that person speaks), and so on indefinitely, both the private domain of competence and the public domain of performance are equally indispensable.

Influential traditions in the discipline of linguistics have embraced an idea that makes little sense, given the fact that language is, after all, just another striking animal behavior. The science of language should begin with fieldwork observation, for performance is ultimately our only evidence for competence. Perhaps the most unfortunate outcome of this idea is that generations of linguists who have eschewed the study of performance now have nothing to say about the essentially social function of language nor about those aspects of social agency, cooperation, and social accountability that universally define our species' unique communicative capacity.

LANGUAGES CONDITION WORLDVIEWS

JOHN McWHORTER
Professor of linguistics, Columbia University; author, The Language Hoax

Since the 1930s, when Benjamin Lee Whorf was mesmerizing audiences with the idea that the Hopi people's language channeled them into a cyclical sense of time, the media and university classrooms have often been abuzz with the idea that the way your language works gives you a particular worldview.

You want this to be true, but it isn't—at least in a way that anyone outside of a psychology laboratory (or academic journal) would be interested in. It's high time thinking people let go of the idea—ever heralded as a possibility but never demonstrated—that different languages represent different ways of experiencing life. Different *cultures* represent different ways of experiencing life, to be sure. And part of a culture is having words and expressions to express it, to be sure. Cell phone. *Inshallah. Feng shui.* But this isn't what Whorfianism (as it's often called) is onto. The idea is that quiet things in a language's structural architecture—how its grammar works, how its vocabulary happens to cut up space—channel how the speaker experiences life.

And in fact psychologists have indeed shown that such things do influence thought—in tiny ways elicitable via fascinatingly peculiar experiments. So, Russian has different words for dark blue and light blue and no one word that just means "blue," and it's been shown that Russians are, indeed,

124 milliseconds faster at matching grades of dark blue to other ones and grades of light blue to other ones. It's also been shown that people whose languages divide nouns into masculine and feminine categories are more likely, if asked, to imagine those things, if they were cartoon characters, talking in the appropriately sexed voice or to associate them with gendered traits.

This kind of thing is neat, but the question is whether the quiet background flutterings of awareness they document can be treated as a worldview. The temptation is endless to suppose that it does. Plus, we're always reminded that no one has said that language *prevents* a speaker from thinking such-and-such but, rather, that language makes it likelier that the speaker *will* think such-and-such.

But we still run up against the fact that languages tell us what we don't want to hear as much as they tell us what's cool (such as Russian blues and tables talking like ladies). Example: in Mandarin Chinese, the same sentence can mean "If you see my sister, you know she's pregnant," or "If you saw my sister, you'd know she's pregnant," or "If you had seen my sister, you'd have known she was pregnant." That is, Chinese leaves hypotheticality to context much more than English does. In the early eighties, psychologist Alfred Bloom, following the Whorfian line, did an experiment suggesting that Chinese makes its speakers somewhat less adept than English speakers at processing hypothetical scenarios.

Whoops! Nobody wanted to hear that. There was a long train of rebuttals, ending in an exhausted draw. But you could do all kinds of experiments that would lead to the same kind of place. Lots of languages in New Guinea have only one word for eating, drinking, and smoking. Does that make them slightly less sensitive to the culinary than other people? Swedish doesn't have a word for "wipe"—you have to erase, take off, and so on. But who's ready to tell the Swedes they don't wipe?

In cases like this, our natural inclination is to say that such things are just accidents and whatever wisp of thought difference an experimenter could elicit based on them has hardly anything to do with what the language's speakers are like, or what their worldview is. But then we have to admit the same thing about the wisps that happen to tickle our fancies.

What creates a worldview is culture—that is, a worldview. And no, it won't work to say that culture and language create a worldview together holistically. Remember, that would mean that Chinese speakers are—holistically—a little dim when it comes to thinking beyond reality.

Who wants to go there? Especially when even starting to go there, decade after decade, has led us down blind alleys? Hopi, it turns out, has plenty of markers of good old-fashioned European-style time. UCLA economist Keith Chen's recent idea that not having a future tense makes a language's speaker more thrifty—pause to wrap your head around that; it's *not* having a future that makes you save money!—has intrigued the media for years now. But if four Slavic languages—Russian, Polish, Czech, and Slovak—have no future tense and yet savings rates in those countries are vastly different, then the whole idea is out the window.

The idea that language is a lens on life should be treated as what it is—something that pans out in terms of quiet results in intense psychological studies but has nothing to do with any humanistic perspective on what it means to be a human being. An awkward aspect of this is that people engaged in trying to document or save the hundreds of languages worldwide threatened with extinction tend to say that the languages must survive because they represent ways of looking at the world. But if they don't, we have to formulate new justifications for those rescue efforts. One hopes that linguists and anthropologists can embrace saving languages simply because they are, in so many ways, magnificent in their own right.

What it comes down to is this: Let's ask how English makes a worldview. Our answer requires that the worldview be shared by Betty White, William McKinley, Amy Winehouse, Jerry Seinfeld, Kanye West, Elizabeth Cady Stanton, Gary Coleman, Virginia Woolf, and Bono. Let's face it, what worldview would that be? Sure, a lab test could likely tease out some infinitesimal squeak of a perceptive predilection shared by all those people. But we wouldn't even begin to think of it as a way of perceiving the world or reflecting a culture. Or if anyone does, then we're on to an entirely new academic paradigm indeed.

THE STANDARD APPROACH TO MEANING

DAN SPERBER

Social and cognitive scientist; research professor emeritus, Centre National de la Recherche Scientifique; director, International Cognition and Culture Institute; coauthor (with Deirdre Wilson), Meaning and Relevance

What is meaning? There are dozens of theories. I suspect, however, that little would be lost if most of them were retired and the others quarantined until we've had a serious conversation as to why we need a theory of meaning in the first place. I'm nominating for retirement just the standard approach to meaning found in the study of language and communication.

There, "meaning" is used to talk about (1) what linguistic items, such as words and sentences, mean, and (2) what speakers mean. Linguistic meanings and speakers' meanings are quite different things. To know a word is to know what its meaning or meanings (if it's ambiguous) are. You acquire this knowledge when you learn to speak a language. You also acquire the ability to construct the meaning of a sentence on the basis of the syntax. The meanings of words and the contribution of the syntax to the meaning of sentences are relatively stable linguistic properties that vary over historical time and across dialects. A speaker's meaning, on the other hand, is a component of an individual intention to modify the beliefs or attitudes of other people through communication.

What justifies, or so it seems, using the same word, "meaning," for these two quite different kinds of phenomena—a linguistic-community-wide stable feature of a language versus an aspect of a social interaction—is a simple and powerful dogma that purports to explain how a speaker manages to convey her meaning to her audience. She does so, we're told, by producing a sentence the linguistic meaning of which matches her speaker's meaning. The job of the addressee then is just to decode.

Alas, this simple and powerful account of how we use linguistic meanings to convey our speakers' meanings is not true. This much is obvious to all students of language. The issue is, How far is it from the truth?

Take an ordinary sentence—say, "She went." Your competence as an English speaker provides you with all the knowledge of that sentence's meaning you need in order to make use of it, either in speaking or in comprehension. This, however, does not come near to telling you what a speaker who utters this sentence on a given occasion might mean. She might mean that Susan Jones had gone home, that the cat had one day left the house and never returned, or that the RMS *Queen Mary 2* had just left the harbor. She might mean that the neighbor carried out her threat to go to the police—or, ironically, that her interlocutor was a fool to imagine that their neighbor would carry out that threat. She might mean, metaphorically, that Nancy Smith had at some point wholly ceased paying attention. And so forth. None of these meanings is fully encoded by the sentence; some aren't even partially encoded. That much is true not just of "She went" but also of the vast majority of English sentences (arguably, of all of them). Linguists and philosophers are aware of this general mismatch between linguistic and speaker's meaning, but most of them treat it as if it were a complication of limited relevance that could be idealized away or left to be investigated by pragmatics, a marginal subfield of linguistics.

The dogma, then, comes with an annotation: The basic coding-decoding mechanism that makes communication possible is quite cumbersome. Using it involves being wholly explicit. Luckily, there's a shortcut: You can avoid the verbosity of full explicitness and rely on your audience to infer rather than decode at least part of your meaning (or all of it, if you use, for instance, a novel metaphor).

There are two problems with this dogma. The first is that the alleged basic mechanism is never used. You never fully encode your meaning. Often, you don't encode it at all. The second problem is that if we can easily infer a speaker's meaning from an utterance that doesn't encode it, then why, in the first place, do we need the alleged basic encoding-decoding mechanism that's so unwieldy?

Imagine a tribe where people who want to go from their valley to the sea always follow a well-trodden path across a low mountain pass. According to the tribe's sages, however, this path is just a shortcut, and the real way (without which there couldn't even have been a shortcut) is a majestic road that goes straight up to the top of the mountain and then straight down to the sea. Nobody has ever seen that road, let alone traveled it, but it has been so much talked about that everybody can visualize it and marvel at the sages' wisdom. Linguistics and philosophy are the home of many such sages.

Most of the time, semanticists start from the dogma I have just criticized. They provide elaborate, often formal analyses of linguistic meanings that match the contents of our conscious thoughts. Are linguistic meanings really like this? Only a minority of researchers is exploring the idea that they might be a very different kind of mental object. Unlike beliefs and intentions, linguistic "meanings" may be just as inaccessible to untutored consciousness as are syntactic properties. They must, on the other hand, be the right kind of objects to serve as input to the unconscious inferences that achieve comprehension.

Pragmaticists and psycholinguists should, for their part, acknowledge that the meanings conveyed by our utterances may be not at all like individual sentences written in our minds in the "language of thought" but rather like partly clear, partly vague reverberating changes in our cognitive environment.

The old dogma that linguistic meanings and speakers' meanings match denies or discounts a blatant gap. This gap is filled by intense cognitive activity of a specifically human kind. Let's retire the dogma and better explore the gap.

THE UNCERTAINTY PRINCIPLE

KAI KRAUSE

Software pioneer; philosopher; author, The History of the Future

It was born out of a mistranslation and has been misused ever since . . . but let us do a little thought experiment first:

Let's say you're a scientist and you notice a phenomenon you'd like to tell the world about. "The brain," you say, "can listen to a conversation and make sense of the frequencies, decode them into symbols and meaning, but when it's confronted with *two* such conversations simultaneously, it cannot deal with both threads in parallel. At best, it can switch back and forth quickly, trying to keep up with the information."

So much for your theory—you formulate your findings and share it with colleagues, it gets argued and debated, just as it should be.

But now something odd happens: Although all your discussions were in English, and you wrote down your idea in English, and a large percentage of leading scientists and Nobel laureates are English-speaking . . . somehow the prevailing language for publication is . . . Mongolian! There's a group in Ulan Bator merrily examining your findings with great interest, and your theory shows up all over the place . . . in Mongolian.

But here's the catch: You wrote that it's not possible to listen to two conversations at the same time and thus their meaning to you is undefined until you decide to follow one of them properly. However, as it turns out, Mongolian has no such word as "undefined"! Instead, it got translated with an entirely different term, "uncertain," and the general interpretation of your theory has suddenly mutated from "one or the other of two conversations will be unknown to you" to the rather distinctly altered interpretation "you can listen to one, but the other will be entirely meaningless." Saying that I am "unable to understand" both conversations properly is one thing, but my inability to perceive it does not render each of the conversations suddenly "meaningless," does it?

All of this is, of course, just an analogy. But it's pretty close to exactly what did happen—just the other way around. The scientist was Werner Heisenberg.

His observation was not about listening to simultaneous conversations but measuring the exact position and momentum of a physical system, which features he described as impossible to determine at the same time. And although he discussed this with numerous colleagues in German (Einstein, Pauli, Schrödinger, Bohr, Lorentz, Born, Planck—just to name some of the Solvay Conference group of 1927), the big step came in the dissemination in English, and there's the Mongolian in our analogy. Heisenberg's idea had quickly been dubbed *Unschärferelation,* which transliterates to "unsharpness relationship"—but as there's really no such term in English ("blurred," "fuzzy," "vague," and "ambiguous" have all been tried), the translation ended up as "the uncertainty principle," when he hadn't used either word at all (some point to Eddington). And what followed is really quite close to the analogy as well: Rather than stating that either position or momentum is "as yet undetermined," it became common usage and popular wisdom to jump to the conclusion that there is complete "uncertainty" at the fundamental level of physics, nature, and even free will and the universe as such. Laplace's Demon was killed as collateral damage (obviously his days were numbered anyway . . .).

Einstein remained skeptical his entire life: To him, the *Unbestimmtheit* (indeterminacy) was on the part of the observer not realizing certain aspects of nature at this stage in our knowledge—rather than proof that nature *itself* is fundamentally undetermined and uncertain. In particular, implications like *Fernwirkung* ("action at a distance") appeared to him *spukhafte* ("spooky," "eerie"). But even in the days of quantum computing, qubits, and tunneling effects, I still would not want to bet against Albert. His intuitive grasp of nature survived many critics, and waves of counterproof ended up counter-counterproved.

And while there is plenty of reason to defend Heisenberg's findings, it's sad to see such a profound meme in popular science based merely on a loose attitude toward translation (there are many other such cases . . .). I would love to encourage writers in French or Swedish or Arabic to point out the idiosyncracies and unique value of those languages—not for semantic pedantry but for the benefit of alternate approaches.

German is good not only for *Fahrvergnügen*, *Weltanschauung,* and *Zeitgeist*. There are many wonderful subtle shades of meaning. It's like a different tool to apply to thinking—and that's a good thing: A great hammer is a terrible saw.

BEWARE OF ARROGANCE! RETIRE NOTHING!

IAN McEWAN

Novelist; author, Amsterdam; On Chesil Beach; Solar; The Children Act

Beware of arrogance! Retire nothing! A great and rich scientific tradition should hang onto everything it has. Truth is not the only measure. There are ways of being wrong that help others to be right. Some are wrong, but brilliantly so. Some are wrong but contribute to method. Some are wrong but help found a discipline. Aristotle ranged over the whole of human knowledge and was wrong about much. But his invention of zoology alone was priceless. Would you cast him aside? You never know when you might need an old idea. It could rise again one day to enhance a perspective the present cannot imagine. It would not be available to us if it were fully retired. Even Darwin in the early 20th century experienced some neglect, until the Modern Synthesis. *The Expression of the Emotions . . .* took longer to be current. William James also languished, as did psychology, once consciousness as a subject was retired from it. Look at the revived fortunes of Thomas Bayes and Adam Smith (especially *The Theory of Moral Sentiments*). We may need to take another look at the long-maligned Descartes. Epigenetics might even restore the reputation of Lamarck. Freud may yet have something to tell us about the unconscious.

Every last serious and systematic speculation about the

world deserves to be preserved. We need to remember how we got to where we are, and we'd like the future not to retire us. Science should look to literature and maintain a vibrant living history as a monument to ingenuity and persistence. We won't retire Shakespeare. Nor should we Bacon.

BIG DATA

GARY MARCUS

Cognitive scientist, New York University; author, Guitar Zero: The New Musician and the Science of Learning

No, I don't literally mean we should stop believing in, or collecting, Big Data. But we should stop pretending that Big Data is magic. There are few fields that wouldn't benefit from large, carefully collected data sets. But lots of people, even scientists, put more stock in Big Data than they should. Sometimes it seems as if half the talk about understanding science these days, from physics to neuroscience, was about Big Data and associated tools like "dimensionality reduction," "neural networks," "machine-learning algorithms," and "information visualization."

Big Data is without a doubt the idea of the moment. Thirty-nine minutes ago as of this writing (according to the Big Data driving Google News), Gordon Moore (for whom Moore's Law is named) "Gave Big to Big Data," MIT debuted an online course for Big Data (forty-four minutes ago), and Big Data was voted strategy+businesses' Strategy of the year. Forbes had an article about Big Data a few hours before that. There were 163,000 hits for a search for big+data+science.

But science still revolves, fundamentally, around a search of the laws that describe our universe. And the one thing Big Data isn't particular good at is—well, identifying laws. Big Data is brilliant at detecting correlation; the more robust your data set, the better chance you have of identifying correlations, even complex ones involving multiple variables. But correlation never was causation and never will be. All the Big Data in the world won't, by itself, tell you whether smoking causes lung cancer. To really understand the relationship between smoking and cancer, you need to run experiments and develop mechanistic understandings of things like carcinogens, oncogenes, and DNA replication. Merely tabulating a massive database of every smoker and nonsmoker in every city in the world, with every detail about when they smoked, where they smoked, how long they lived, and how they died would not, no matter how many terabytes it occupied, be enough to induce all the complex underlying biological machinery.

If it makes me nervous when people in the business world put too much faith in Big Data, it makes me even more nervous to see scientists do the same. Certain corners of neuroscience have taken on an "If we build it, they will come" attitude, presuming that neuroscience will sort itself out as soon as we have enough data.

It won't. If we have good hypotheses, we can test them with Big Data, but Big Data shouldn't be our first port of call; it should be where we go once we know what we're looking for.

THE STRATIGRAPHIC COLUMN

CHRISTINE FINN

Archaeologist, journalist; author, Artifacts: An Archaeologist's Year in Silicon Valley

Digging for the past has timed out. Digerati are the gatherers now. The law of stratigraphy has held well for archaeology as a means and a concept—the vertical quest exposing time's layers to be read like a book of changes. The exactitude associated with the act of going down, with that of going back and understanding human behavior through geology. The Victorians took up barrow-digging and brought the old stuff home as souvenirs of a Sunday pursuit.

Then archaeologists called it a science, employed the same tools as grave-diggers—spades, buckets—descended six feet under, and brought exactitude to the trenches. But even Schliemann's 19th-century tunneling through layers of dull—to him—prehistory in search of gold was in some ways a prelude to what we have now, exposure to an accumulation of relative yesterdays.

We cherry-pick the past. Time-zone concerns are so over. Blogs are a hoard of content, only as fresh as the day they're retrieved. Archive photos and just-taken selfies get uploaded together onto timelines that run laterally. Half-forgotten news hangs around the Internet, and it surfaces—that old-school term again—as new news to the fresh viewer.

So what is fieldwork now? Look to the new(ish) field of contemporary archaeology, which has its "excavators" channeling anthropology. These are surface workers, seeing escalating and myriad rates of change as lateral observations that connect a series of presents, which oscillate and merge new and old. No hands get dirty in this type of dig. But what is dug up tends to linger under the fingers.

THE HABITABLE-ZONE CONCEPT

DIMITAR D. SASSELOV

Professor of astronomy, Harvard University; director, Harvard Origins of Life Initiative; author, The Life of Super-Earths

"The habitable zone" defines those distances from a star where an Earth-like planet would have the surface temperatures for water to be liquid. In our solar system, this zone stretches from midway between the orbits of Venus and Earth out to Mars. Its boundaries are approximate, applied to different planetary systems, and sometimes the concept applies more broadly—for example, to our galaxy. The habitable-zone concept has a venerable history in the search for life beyond Earth; most recently, it has contributed to the spectacular success of NASA's *Kepler* exoplanet-hunting mission. However, in the post–*Kepler* era it's a scientific concept ready for retirement.

The simple definition of the zone is appealing to use in statistical estimates of habitable planets, because it depends on just a few parameters that are easy to measure. The idea is also easy to grasp: not too hot, not too cold—the Goldilocks zone. Simple and robust statistics are crucial to estimating the abundance and distribution of small, rocky planets like Earth in the galaxy, and the *Kepler* space mission excelled at that. If our goal now is to search for life, then it's good to know where we should be heading—to habitable exoplanets. But the "habitable" in the habitable zone is a misnomer, or at least a gross overstatement, since it isn't the zone per se that is habitable but certain planetary environments, which may even exist outside it. In our own solar system, we contemplate alien life beyond the zone's confines, on certain moons of Jupiter and Saturn. Today we need a concept of what makes an environment habitable—capable of letting life emerge and of sustaining it over geological time scales, be it on a planet or a moon. Finding out what makes a planet "living" and how to recognize a living planet with our telescopes is the big question.

The year 2013 was a historic one in the search for alien life. Thanks to *Kepler* and other exoplanet surveys, we now know that Earth-like planets are so common that many close analogs to our home planet should reside in our neighborhood of the galaxy. This makes them amenable to remote-sensing exploration with existing technology and telescopes under construction. The search for life is set to begin, but we need to understand better what to look for.

In retiring the habitable-zone concept, it makes sense to revert to its original mid-20th-century name—the liquid-water belt, a region important to the rich geochemistry of rocky planets. Living planets there will feel like home to us.

ROBOT COMPANIONS

SHERRY TURKLE

Abby Rockefeller Mauzé Professor of the Social Studies of Science and Technology, MIT; author, Alone Together

In the early 1980s, I interviewed a young student of Marvin Minsky's, one of the founders of artificial intelligence. The student told me that as he saw it, his hero, Minsky, was trying to build a machine beautiful enough that "a soul would want to live in it." More recently, we are perhaps less metaphysical, more practical. We envisage elder-care-bots, nanny-bots, teacher-bots, sex-bots. To go back to Minsky's student, these days we're trying to invent machines not that souls would want to live in but that we would want to live with. We are trying to invent machines that a self would want to love.

The dream of the artificial confidante and then love object confuses categories best left unmuddled. Human beings have bodies and a life cycle, live in families, and grow up from dependence to independence. This gives them experiences of attachment, loss, pain, fear of illness—and of course the experience of death—that are specific and that we don't share with machines. I don't mean that machines can't get smart or learn a stunning amount of things—more things, certainly, than people can know. But they're the wrong object for the job when we want companionship and love.

A machine companion for instrumental help (to keep you safe in your home, to help with cleaning or with reaching high shelves) is an excellent idea. A machine companion for conversation about human relationships seems a bad one. A conversation about human relationships is species-specific. These conversations depend on both parties having the experiences that come from having a human body, human limitations, a human life cycle.

I see us embarked on a voyage of forgetting.

We forget about the care and conversation that can pass only between people. The word "conversation" derives from words that mean to tend to each other, to lean toward each other. To converse, you have to listen to someone else, put yourself in their place, read their body, their voice, their tone, their silences. You bring your concern and experience to bear, and you expect the same. A robot that shares information is an excellent project. But if the project is companionship and mutuality of attachment, you want to lean toward a human.

When we think, for example, about giving children robot babysitters, we forget that what makes children thrive is learning that people care for them in a stable and consistent way. When children are with people, they recognize how the movement and meaning of speech, voice, inflection, faces, and bodies flow together. Children learn how human emotions play in layers, seamlessly and fluidly. No robot has this to teach.

There's a general pattern in our discussions of robot companionship: I call it "From better than nothing to better than anything." I hear people begin with the idea that robot companionship is better than nothing, as in "There are no people for these jobs"—for example, jobs in nursing homes or as babysitters. And then they start to exalt the possibilities of what simulation can offer. In time, people start to talk as though what we would get from the artificial might be better than what life could provide. Child-care workers might be abusive.

Nurses might make mistakes; nursing home attendants might not be clever or well educated.

The appeal of robotic companions carries our anxieties about people. We see artificial intelligence as a risk-free way to avoid being alone. We fear we won't be there to care for each other. We're drawn to the robotic because it offers the illusion of companionship without the demands of friendship. Increasingly, people even suggest that it might offer the illusion of love without the demands of intimacy. We're willing to put robots in places where they have no place, not because they belong there but because of our disappointments with each other.

For a long time, putting hope in artificial intelligence or robots has expressed an enduring technological optimism—a belief that as things go wrong, science will go right. In a complicated world, robots have always seemed like calling in the cavalry. Robots save lives in war zones, in operating rooms; they can function in deep space, in the desert, in the sea, wherever the human body would be in danger. But in the pursuit of artificial companionship, we're not looking for the feats of the cavalry but the benefits of simple salvations.

What are the simple salvations? These are the hopes that artificial intelligences will be our companions. That talking with us will be their vocation. That we'll take comfort in their company and conversation.

In my research over the past fifteen years, I've watched these hopes for the simple salvations persist and grow stronger, even though most people don't have experience with an artificial companion at all but with something like Siri, Apple's digital assistant on the iPhone, where the conversation is most likely to be "Locate a restaurant" or "Locate a friend." But what my research shows is that even telling Siri to "locate a friend" moves quickly to the fantasy of finding a friend in Siri, something like a best friend, but in some ways better: one you can always talk to, one that will never be angry, one you can never disappoint.

When people talk this way, about friendship without mutuality, about friendship on tap, the simple salvations of artificial companionship don't seem so simple to me. For the idea of artificial companionship to become our new normal, we have to change ourselves, and in the process we're remaking human values and human connection. We change ourselves even before we make the machines. We think we're making new machines, but really we're remaking people.

"ARTIFICIAL INTELLIGENCE"

ROGER SCHANK

Artificial-intelligence theorist; executive director, Engines for Education; author, Teaching Minds

It was always a terrible name, but it was also a bad idea. Bad ideas come and go, but this particular idea—that we would build machines that are just like people—has captivated popular culture for a long time. Nearly every year, a new movie, or a new book, appears with a new kind of robot that's just like a person. But that robot will never appear in reality. It's not that artificial intelligence has failed; no one actually ever tried. (There, I said it!)

Oxford physicist David Deutsch has said, "No brain on Earth is yet close to knowing what brains do. . . . The enterprise of achieving it artificially—the field of 'artificial intelligence' . . . has made no progress whatever during the entire six decades of its existence."[s] He adds that he thinks machines that think like people will happen some day.

Let me put that remark in a different light. Will we eventually have machines that feel emotions like people? When that question is asked of someone in AI, the response might be about how we could get a computer to laugh or cry or be angry. But actually feeling?

Or let's talk about learning. A computer can learn, can't it? That's artificial intelligence right there. No machine would be smart if it couldn't learn, but does the fact that machine learning has enabled the creation of a computer that can play Jeopardy or provide data about purchasing habits of consumers mean that AI is on its way?

The fact is that the name "AI" made outsiders to AI imagine goals for AI that AI never had. The founders of AI (with the exception of Marvin Minsky) were obsessed with chess playing and problem solving (the Tower of Hanoi problem was a big one). A machine that plays chess well does just that; it isn't thinking, nor is it smart. It certainly isn't acting like a human. The chess-playing computer won't play worse one day because it drank too much the night before or had a fight with its wife.

Why does this matter? Because a field that started out with a goal different from what its goal was perceived to be is headed for trouble. The founders of AI, and those who work on AI still (me included), want to make computers do things they cannot now do in the hope that something will be learned from this effort or that something will have been created that's of use. A computer that can hold an intelligent conversation with you would be potentially useful. I'm working on a program now that will hold an intelligent conversation about medical issues with a user. Is my program intelligent? No. The program has no self-knowledge. It doesn't know what it's saying, and it doesn't know what it knows. The fact that we've stuck ourselves with this silly idea of intelligent machines, or AI, causes people to misperceive the real issues.

I declare "artificial intelligence" dead. The field should be renamed "the attempt to get computers to do really cool stuff," but of course it won't be. You'll never have a friendly household robot with whom you can have deep, meaningful conversations. I happened to be a judge at last year's Turing Test (known as the Loebner Prize). The stupid stuff that was supposed to be AI was just that—stupid. It took maybe 30 seconds to figure which was a human and which was a computer.

People don't just get fed knowledge. I've raised a couple of people myself. I fed them food, not knowledge. I answered their questions, but those were their own self-generated questions. I tried to help them get want they wanted, but it was (and is) their own deeply felt wants I was dealing with. Humans are born with individual personalties and their own set of wants and needs, and they express them early on. No computer starts out knowing nothing and gradually improves by interacting with people. We always kick that idea around when we talk about AI, but no one ever does it, because it really isn't possible. Nor should it be the goal of the field formerly known as AI. The goal should be figuring out what great stuff people do and seeing if machines can do bits and pieces of that. A chess-playing computer is nice to have, I suppose, but it won't tell you much about how people think, nor will it suddenly get interested in learning a new game to play because it's bored with chess.

There really is no need to create artificial humans anyway. We have enough real ones already.

THE MIND IS JUST THE BRAIN

TANIA LOMBROZO

Associate professor of cognitive psychology,
University of California, Berkeley

In the beginning, there was dualism. Descartes famously posited two kinds of substance, nonphysical mind and material body. Leibniz differentiated mental and physical realms. But dualism faced a challenge—explaining how mind and body interact. The mind executes an intention to raise a finger and, behold, it rises! The body brushes against something sharp, and the mind registers pain.

We now know, of course, that mind and brain are intimately connected. Injuries to the brain can alter perceptual experience, cognitive abilities, and personality. Changes in brain chemistry can do the same. There's no "mental substance" that appears along some phylogenetic branch of our evolutionary history, nor a point in ontogeny during which we receive a nonphysical infusion of mind-stuff. We've come a long way from Ambrose Bierce's formulation of the mind in *The Devil's Dictionary* as "a mysterious form of matter secreted by the brain."

In fact, it appears the mind is just the brain. Or perhaps, to quote Marvin Minsky, "Minds are simply what brains do." If we want to understand the mind, we should look to neuroscience and the brain for the real answers.

Or maybe not.

In our enthusiasm to find a scientifically acceptable alternative to dualism, some of us have gone too far the other way, adopting a stark reductionism. Understanding the mind is not just a matter of understanding the brain. But then, what is it a matter of? Many alternatives to the *mind* = *brain* equation seem counterintuitive or spooky. Some suggest that the mind extends beyond the brain to encompass the whole body, or even parts of the environment, or that the mind is not subject to the laws of physics.

Are there other options? Indeed there are. But given that mind and brain are pretty heady matters (so to speak), it helps to think about a more concrete—and tastier—example. Consider baking.

I'm an antireductionist about baking. It's not that I believe in a "cake substance" materially distinct from flour and sugar and leavening. And it's not that I think cakes have some magical metaphysical property (though the best ones sort of do). The tenets of baking antireductionism are far less controversial, and they stem from what we want our "theory of baking" to provide. We want to understand why some cakes turn out better than others, and what we can do to achieve better baked goods in the future. Should we change an ingredient? Mix the batter less vigorously?

Answering these questions can appeal to chemistry and physics. But a theory of baking wouldn't be very useful formulated in terms of molecules and atoms. As bakers, we want to understand the relationship between (for example) mixing and texture, not between kinetic energy and protein hydration. The relationships between the variables we can tweak and the outcomes we care about happen to be mediated by chemistry and physics, but it would be a mistake to adopt "cake reductionism" and replace the study of baking with the study of physical and chemical interactions among cake components.

Of course, you could decide you're not interested in baking and thus reject the theoretical constructs of my baking theory in favor of chemistry and physics. But if you *are* interested in the project of explaining, predicting, and controlling the quality of your baked goods, you'll need something like a baking theory to work with.

Now consider the mind. Most of us are interested in a theory of the mind because we want to explain, predict, and control behaviors, mental states, and experiences. Given that mental phenomena are physically realized in the brain just as cake properties are physically realized by their ingredients and interactions, understanding the brain is incredibly useful. But if we want to know (for instance) how to influence minds to achieve particular behaviors, it would be a mistake to look for explanations solely at the level of the brain.

These reflections won't be news to many philosophers, but they're worth repeating. Rejecting the mind in an effort to achieve scientific legitimacy—a trend we have seen both in behaviorism and some popular manifestations of neuroscience—is unnecessary and unresponsive to the aims of scientific psychology. Understanding the mind isn't the same as understanding the brain. Fortunately, though, we can achieve such understanding without abandoning scientific rigor. Or, to adopt another baking analogy, we can have our cake and eat it, too.

MIND VERSUS MATTER

FRANK WILCZEK

Herman Feshbach Professor of Physics, MIT; corecipient, 2004 Nobel Prize in physics; author, The Lightness of Being

The distinction between mind and matter is embedded in everyday language and thinking, and even more deeply in philosophy and theology. The great philosopher/theologian George Berkeley, who famously grounded matter in the mind of God, summed it up in a witticism:

What is mind? No matter.
What is matter? Never mind.

Science has long found it useful to accept this duality, as a methodology if not as a doctrine. In modern physics, matter obeys its own mathematical laws, independent of what anyone—even, or maybe especially, God—thinks. But the distinction is doomed, and its passing will change our view of everything—everything, that is, which is mind/matter. Already the walls of separation are crumbling. Three developments have irreversibly undermined them:

(1) We've learned what matter is. And our new matter, informed over the course of the 20th century by the revelations of relativity, quantum mechanics, and transformational symmetry, is far stranger and richer in potential than anything our ancestors could have dreamed of. It can dance in intricate, dynamic patterns;

it can exploit environmental resources to self-organize and export entropy.

(2) We've learned—theoretically, through Turing's vision, and practically, through the rise of ubiquitous computing—that many accomplishments once viewed as prerogatives of mind, from playing chess to planning itineraries to suggesting friends and sharing interests, are things that machines (whose design hides no secrets), by pure computation, can do quite well.

(3) We've learned a lot about how the human mind works, as a special capacity of matter. We now know that many aspects of perception begin as specific molecular events. Great challenges remain to bring understanding of memory, emotion, and ultimately creative thought to the same level, but there's every reason to think they, too, will come into focus. At least, no showstoppers have yet appeared.

The eternal, ever vague "problems" of free will and consciousness will be retired, with due respect, when mechanistic understanding of how human minds actually work brings in more powerful, less nebulous concepts (as has already happened for computation).

More interesting is the question of consequences. Here's a relevant thought experiment: Imagine an artificial intelligence with humanlike insight, contemplating her own blueprint. What would she make of it? I think it's overwhelmingly likely that among her first thoughts would be how to begin making improvements. This processor could be faster, that memory more capacious—and, above all, the reward system more rewarding!

Our heroine would surely be inspired, as I am, by William Blake's prophecy: "If the doors of perception were cleansed, everything would appear to man as it is, Infinite."

In bad science fiction, androids are sometimes horrified to learn they're "mere machines." Following the instruction of the Delphic oracle to "Know thyself," we find ourselves making a similar discovery. The wise and mature reaction to the realization that mind and matter are mind/matter is to take joy in what a wonderful thing mind/matter can be, and is.

INTELLIGENCE AS A PROPERTY

ALEXANDER WISSNER-GROSS

Scientist; inventor; entrepreneur; Institute Fellow, Institute for Applied Computational Science, Harvard University

Since long before Erwin Schrödinger's seminal 1944 work, *What Is Life?*, physicists have aspired to rigorously define the characteristics that distinguish some matter as living and other matter as not. However, the analogous task of identifying the universally distinguishing physical properties of intelligence has remained largely underappreciated.

Based on recent discoveries, I have come to suspect that the reason for this lack of progress in physically defining intelligence is due to the scientific concept of intelligence as a static property rather than a dynamical process—a concept ready for retirement.

In particular, recent results have shown that a rudimentary physical process called "causal entropic forcing" can replicate model versions of signature cognitive adaptive behaviors seen previously only in humans and certain nonhuman animal intelligence tests. These findings suggest that a variety of key characteristics associated with human intelligence, including upright walking, tool use, and social cooperation, should instead be viewed as side effects of a deeper dynamical process that attempts to maximize future freedom of action. This freedom-maximizing process can be said to meaningfully exist only over an extended time period and therefore is not a static property.

It's time we retired studying intelligence as a property.

THE GRAND ANALOGY

DAVID GELERNTER

Computer scientist, Yale University; chief scientist, Mirror Worlds Technologies; author, America-Lite: How Imperial Academia Dismantled our Culture (and ushered in the Obamacrats)

Today computationalists and cognitive scientists—those researchers who see digital computing as a model for human thought and the mind—are nearly unanimous in believing the Grand Analogy and teaching it to their students. And whether you accept it or not, the analogy is a milestone of modern intellectual history. It partly explains why a solid majority of contemporary computationalists and cognitive scientists believe that eventually you'll be able to give your laptop a *real* (not simulated) mind by downloading and executing the right software app. Whereupon if you tell the machine, "Imagine a rose," it will conjure one up in its mind, just as you do. Tell it to "recall an embarrassing moment," and it will recall something and feel embarrassed, just as you might. In this view, embarrassed computers are just around the corner.

But no such software will ever exist, and the analogy is false and has slowed our progress in grasping the actual phenomenology of mind. We've barely begun to understand the mind from inside. But what's wrong with this suggestive, provocative analogy? My first reason is old; the other three are new.

(1) The software-computer system relates to the world in a fundamentally different way from the mind-brain system. Software moves easily among digital computers, but each human mind is (so far) wedded permanently to one brain. The rela-

tionship between software and the world at large is arbitrary, determined by the programmer; the relationship between mind and world is an expression of personality and human nature, and no one can rearrange it.

There are computers without software, but no brains without minds. Software is transparent. I can read off the precise state of the entire program at any time. Minds are opaque—there's no way I can know what you're thinking unless you tell me. Computers can be erased; minds cannot. Computers can be made to operate precisely as we choose; minds cannot. And so on. Everywhere we look we see fundamental differences.

(2) The Grand Analogy presupposes that minds are machines, or virtual machines. But a mind has two equally important functions: *doing* and *being*. A machine is only for *doing*. We build machines to act for us. Minds are different: Yours might be wholly quiet, *doing* ("computing") nothing—yet you might be feeling miserable or exalted, or you might merely *be* conscious.

Emotions, in particular, aren't actions, they're ways *to be*. And emotions—states of being—play an important part in the mind's cognitive work. They allow you, for instance, to feel your way to a cognitive goal. ("He walked to the window to recollect himself, and feel how he ought to behave." Jane Austen, *Persuasion*.) Thoughts contain information, but feelings (mild wistfulness, say, on a warm summer morning) contain none. Wistfulness is merely a way to *be*.

Until we understand how to make digital computers feel (or experience phenomenal consciousness), we have no business talking up a supposed analogy between *mind:brain* and *software:computer*. Those who note that computers-that-can-feel are incredible are sometimes told, "You assert that many billions of tiny, meaningless computer instructions, each unable to feel, could never create a *system* that feels. Yet neurons are also tiny, 'meaningless,' and feel nothing—but 100 billion of those yields a brain that *does* feel." Which is irrelevant; 100 billion neurons yield a brain that supports a mind, but 100 billion sand grains or used tires yields nothing. You need billions of the right *article* arranged in the right way to get feeling.

(3) The process of growing up is innate to the idea of human being. Social interactions and body structure change over time, and the two sets of changes are intimately connected. A toddler who can walk is treated differently from an infant who can't. No robot could acquire a humanlike mind unless it could grow and change physically, interacting with society as it did. But even if we focus on static, snapshot minds, a human mind requires a human body. Bodily sensations create mind states that cause physical changes that create further mind changes. A feedback loop. You're embarrassed; you blush; feeling yourself blush, your embarrassment increases. Your blush deepens. We don't think with our brains only. We think with our brains and bodies together. We might build simulated bodies out of software, but simulated bodies can't interact in human ways with human beings. And we must interact with other people to become thinking persons.

(4) Software is inherently recursive; recursive structure is innate to the idea of software. The mind is not and cannot be recursive. A recursive structure incorporates smaller versions of itself—an electronic circuit made of smaller circuits, an algebraic expression built of smaller expressions. Software is a digital computer realized by another digital computer. (You can find plenty of definitions of "digital computer.") "Realized by" means made-real-by or embodied-by. The software

you build is capable of exactly the same computations as the hardware on which it executes. Hardware is a digital computer realized by electronics (or some equivalent medium). Suppose you design a digital computer; you embody it using electronics. So you've got an ordinary computer, with no software. Now you design another digital computer—an operating system, like Unix. Unix has a distinctive interface and, ultimately, the exact same computing power as the machine it runs on. You run your new computer (Unix) on your hardware computer. Now you build a word processor (yet another dressed-up digital computer) to run on Unix. And so on, *ad infinitum*. The same structure (a digital computer) keeps recurring. Software is inherently recursive. The mind is not and cannot be. You cannot "run" another mind on yours, and a third mind on that, and a fourth atop the third.

Much has been gained by mind science's obsession with computing. Computation has been a useful lens to focus scientific and philosophic thinking on the essence of mind. The last generation has seen, for example, a much clearer view of the nature of consciousness. But we've always known ourselves poorly. We still do. Your mind is a room with a view, and we still know the view (objective reality) a lot better than the room (subjective reality). Today, subjectivism is reemerging among those who see through the Grand Analogy. Computers are fine, but it's time to return to the mind itself and stop pretending we have computers for brains. We'd be unfeeling, unconscious zombies if we did.

GRANDMOTHER CELLS

TERRENCE J. SEJNOWSKI

Computational neuroscientist; Francis Crick Professor, Salk Institute for Biological Studies; coauthor (with Patricia S. Churchland), The Computational Brain

In 2004, an epilepsy patient at the UCLA Medical Center whose brain was being monitored to detect the origin of the seizures was shown a series of pictures of celebrities. Electrodes implanted in the memory centers of the patient's brain reported spikes in response to the photos. One of the neurons responded vigorously to several pictures of Jennifer Aniston but not to other famous people. A neuron in another patient would respond only to pictures of Halle Berry, and even to her name, but not to pictures of Bill Clinton or Julia Roberts or the names of other famous people.

Such cells had been predicted fifty years ago, when it first became possible to record from single neurons in the brains of cats and monkeys. It was thought that in the hierarchy of visual areas of the cerebral cortex, the response properties of the neurons became more and more specific the higher the neuron was in the hierarchy—perhaps so specific that a single neuron would respond only to pictures of a single person. This became known as the grandmother-cell hypothesis, after the putative neuron in your brain that recognizes your grandmother. The team at UCLA seemed to have found such cells. Single neurons were also found that recognized specific objects and buildings, like the Sydney Opera House.

Despite this striking evidence, the grandmother-cell hypothesis is unlikely to be correct, or even a good explanation for these recordings. We're beginning to collect recordings from hundreds of cells simultaneously in mice, monkeys, and humans, and these are leading to a different theory for how the cortex perceives and decides. Nonetheless, the grandmother-cell hypothesis continues to have adherents, and the thinking that derives from focusing on single neurons still permeates the field of cortical electrophysiology. We'd make progress more quickly if we could retire the proverbial grandmother cell.

According to the grandmother-cell hypothesis, you perceive your grandmother when the cell is active, so it shouldn't fire to any other stimulus. Only a few hundred pictures were tested, and many more pictures were not tested, so we really don't know how selective the Jennifer Aniston cell was. Second, the likelihood that the electrode by chance happened to record from the only Jennifer Aniston neuron in the brain is low; it's more likely that there are many thousands. The same for the Halle Berry neuron, for everyone you know and every object you can recognize. There are many neurons in the brain, but not enough for each object and name that you know. An even deeper reason to be skeptical of the grandmother-cell hypothesis is that the function of a sensory neuron is only partially determined by its response to sensory inputs. Equally important is the output of the neuron and its effects downstream on behavior.

In monkeys, where it has been possible to record from many neurons simultaneously, stimuli and task-dependent signals are broadly distributed over large populations of neurons, each tuned to a different combination of features of the stimuli and task detail. The properties of such distributed representations were first studied in artificial neural networks in the 1980s.

Populations of simple model neurons called "hidden units" were trained to perform a mapping between a set of input units and a set of output units; these hidden units developed patterns of activity for each input that was highly distributed, like what has been observed in populations of cortical neurons. For example, the input units could represent faces from many different angles, and the output units could represent the names of the people. After being trained on many examples, each of the hidden units coded different combinations of features of the input units, such as fragments of eyes, noses, or head shapes.

A distributed representation can be used to recognize many versions of the same object, and the same set of neurons can recognize many different objects by differentially weighting their outputs. Moreover, the network can generalize by correctly classifying new inputs from outside the training set. Much more powerful versions of these early neural-network models, with more than twelve layers of hidden units in a hierarchy like that in our visual cortex, and using deep learning to adjust billions of synaptic weights, are now able to recognize tens of thousands of objects in images. This is a breakthrough in artificial intelligence, because performance continues to improve as the size of the network and the number of training examples increases. Companies worldwide are racing to build special-purpose hardware that would scale up these architectures. There's still a long way to go before the current systems approach the capacity of the human brain, which has a billion synapses in every cubic millimeter of cortex.

How many neurons are needed in a population that can discriminate between many similar objects, such as faces? From imaging studies, we know that many areas of the brain respond to faces, some with a high degree of selectivity. We'll need to sample many neurons widely from these areas. The answer to this question may be a surprise, because there are also sound theoretical arguments for minimizing the numbers of neurons in the representation of an object. First, sparse coding would be more energy efficient. Second, learning a new object in the same population of neurons leads to interference with the others being represented in the population. An effective and efficient representation would be sparsely distributed.

In ten years, 1,000 times more neurons will be recorded and manipulated than is now possible, and new techniques are being developed to analyze them, which could lead to a deeper understanding of how activity in populations of neurons gives rise to thoughts, emotions, plans, and decisions. We may soon know the answer to the question of how many neurons represent an object or a concept in our brain—but will this retire the grandmother-cell hypothesis?

BRAIN MODULES

PATRICIA S. CHURCHLAND

Philosopher, neuroscientist, University of California, San Diego; author, Touching a Nerve

The concept of "module" in neuroscience (meaning *sufficient for a function, given gas-in-the-tank background conditions*) invariably causes more confusion than clarity. The problem is that any neuronal business of any significant complexity is underpinned by spatially distributed networks, and not just incidentally but essentially—and not just cortically but between cortical and subcortical networks. This is true, for example, of motion perception and pattern recognition, as well as motor control and reinforcement learning, not to mention feelings such as mustering courage to face a threat or deciding to hide instead of run. It's true of self-control and moral judgment. It's likely to be true of conscious experience. The output of a network can vary as the activity of the network's individual neurons varies. What's poorly understood is how nervous systems solve the coordination problem—that is, how does the brain orchestrate the right pattern of neuronal activation across networks to get the job done?

This isn't all that's amiss with "module." Traditionally, modules are supposed to be encapsulated—aka insulated. But even the degree to which an early sensory area such as primary visual cortex (V1) is encapsulated has been challenged. Visual neurons in V1 double their firing rate if the animal is running, no matter the identity of the visual input across conditions (to take but one example). To add to the "module" mess, it turns out that specialization in an area such as the V1 appears to depend to some nontrivial degree on the statistics of the input. Visual cortex is visual largely because it's connected to the retina and not the cochlea, for example. Notice that in blind subjects, the visual cortex is recruited in reading braille, a high-resolution spatial—and somatosensory—task. As would be expected if specialization depends on the statistics of the network's input, infant brains have much more plasticity in regional specialization than do mature brains. Doris Trauner and Elizabeth Bates discovered that human infants with a left hemispherectomy can learn language quite normally, whereas an adult who undergoes the same surgery will have severe language deficits.

I think of "module" in the way I think of "nervous breakdown"—mildly useful in the old days when we had no clue about what was going on under the skull, but of doubtful explanatory significance these days.

BIAS IS ALWAYS BAD

TOM GRIFFITHS

Associate professor of psychology; director, Computational Cognitive Science Laboratory and the Institute of Cognitive and Brain Sciences, University of California, Berkeley

Being biased seems like a bad thing. Intuitively, we equate rationality and objectivity. When faced with a difficult question, a rational agent, it seems, shouldn't have a predisposition for one answer over another. If a new algorithm designed to find objects in images or interpret natural language is described as biased, it sounds like a poor algorithm. And when psychology experiments show that people are systematically biased in the judgments they form and the decisions they make, we begin to question human rationality.

But bias isn't always bad. In fact, for certain kinds of questions the only way to produce better answers is to be biased.

Many of the most challenging problems humans solve are known as inductive problems—problems where the right answer cannot be definitively identified based on the available evidence. Finding objects in images and interpreting natural language are two classic examples. An image is just a two-dimensional array of pixels—a set of numbers indicating whether locations are light or dark, green or blue. An object is a three-dimensional form, and many different combinations of three-dimensional forms can result in the same pattern of numbers in a set of pixels. Seeing a particular pattern of numbers doesn't tell us which of these possible three-dimensional forms is present: We have to weigh the available evidence and make a guess. Likewise, extracting the words from the raw sound pattern of human speech requires making an informed guess about the particular sentence a person might have uttered.

The only way to solve inductive problems well is to be biased. Because the available evidence isn't enough to determine the right answer, you need to have predispositions independent of that evidence. And how well you solve the problem—how often your guesses are correct—depends on having biases that reflect how likely different answers are.

Human beings are very good at solving inductive problems. Finding objects in images and interpreting natural language are two problems that people still solve better than computers. And the reason is that human minds have biases finely tuned for solving these problems.

The biases of the human visual system are apparent in many visual illusions—images that result in a surprising discrepancy between our biased guesses and what's actually in the world. The rarity of visual illusions in real life is testimony to the utility of those biases. By studying the kinds of illusions the human visual system is susceptible to, we can identify the biases that guide perception and instantiate those biases in algorithms used by computers.

Human biases in interpreting language are demonstrated in the game of Telephone, or when we misinterpret the lyrics of a song. It's also easy to discover the biases built into speech-recognition software. I once left my office for a meeting, locking the door behind me, and came back to find that a stranger had broken in and typed a series of poetic sentences into my computer. Who was this person, and what did the message mean? After a few spooky, puzzling minutes, I realized that I'd left my speech-recognition software running and the sentences were the guesses it had produced about what the rustling of the trees outside my window meant. But the fact that they were fairly intelligible English sentences reflected the biases of the software, which didn't even consider the possibility that it was listening to the wind rather than a person.

Things that people do well—vision and language—depend heavily on our being biased toward particular answers. Algorithms that solve those problems well have similar biases. So we shouldn't be surprised to discover that people are systematically biased in other domains. These biases don't necessarily reflect a deviation from rationality; they reflect the difficulty of the problems that humans need to solve. And one way to make computers better at solving these problems is understanding exactly what human biases are like for different problems.

In arguing that bias isn't always bad, I'm not claiming that it's always good. Objectivity can be an ideal we strive for on moral grounds—say, when assessing other people. The more information and time we have available, the closer we can get to this ideal. But this kind of objectivity is a luxury, at odds with reaching the right answers in limited time from small amounts of evidence. When solving inductive problems, it can be rational to be biased.

CARTESIAN HYDRAULICISM

ROBERT KURZBAN

Professor of psychology, University of Pennsylvania; director, Penn Laboratory for Experimental Evolutionary Psychology (PLEEP); author, Why Everyone (Else) Is a Hypocrite

In the 17th century, René Descartes proposed that the nervous system worked a bit like the nifty statues in the royal gardens of Saint-Germain, whose moving parts were animated by water running through pipes inside them. Descartes' idea is illustrated in the well-known line drawing in many introductory psychology textbooks, showing a person sticking his foot in a fire, presumably to illustrate Descartes' idea about hydraulic reflexes.

Three centuries on, in the mid-1900s, the detritus of the hydraulic conception of behavior, now known to be luminously wrong, was strewn about here and there. In the scholarly literature, for instance, there were traces in Freud's corpus—catharsis will relieve all that *pressure.* Among the Folk, hydraulic metaphors were—and still are—used to express mental states: I'm going to *blow my top.* Having written an essay for *Edge* today, I feel *drained.*

There is, to be sure, still plenty of debate about how the mind works. No doubt even in the answers to this year's *Edge* Question there will be spirited discussion about how well the brain-as-device-that-computes notion is doing in attempts to advance psychology. Still, while the computational theory of mind may not have won over everyone, the hydraulic model that Descartes proposed is dead and buried.

Well, dead anyway. Buried . . . maybe not. And, to be sure, hydraulics is, as it turned out, the right explanation for a pretty important (male) biological function—just not the one Descartes had in mind. The metaphors recruiting the intuition that the mind is built of fluid-filled pipes along with junctions, valves, and reservoirs point to the possibility that Descartes was drawn to the notion of a hydraulic mind not only because of the technology of the day but also because there's something intuitively compelling about the idea.

And, indeed, Cartesian hydraulics has been revived in at least one incarnation in the scholarly literature, though I doubt it's the only one. For the last decade or so, some researchers have been advancing the notion that there is a "reservoir" of willpower. You can't have an empty reservoir, the theory goes, in order to exert self-control—resisting eating marshmallows, avoiding distractions, etc.—and as the reservoir gets *drained*, it become harder and harder to exert self-control.

Given how wrong Descartes was about how the mind works, it's pretty clear that this sort of idea just can't be right. There have recently been a number of experimental results that disconfirm predictions made by the model, but that's not why the idea should be abandoned. Or, at least, the data aren't the best reason the idea should be abandoned. The reason the idea should be left to die is the same reason that Descartes' idea should be: Although the mind might not work just like a digital computer—no doubt the mind is different from your basic PC in any number of important ways—we do know that computation of some sort is much, much more likely to be a good explanation for human behavior than hydraulics is.

People will disagree about whether Max Planck was right about the speed of scientific change. Psychology, I would argue, has a couple of handicaps that might make the discipline more susceptible to Planck's worries than some other disciplines.

First, theories in psychology are often driven by—indeed, held captive by—our intuitions. I'm fond of the way Daniel Dennett put it in *Consciousness Explained* (1991), when he was talking about the (also luminously wrong) idea of the Cartesian Theater—the dualist idea that there's "a special center in the brain," the epicenter of identity, the One and True Me, the wizard behind the curtain. He thought this notion was "the most tenacious bad idea bedeviling our attempts to think about consciousness." Human intuitions tell us that there's a special "me" in there somewhere, an intuition that serves to resurrect the idea of a special center over and over again.

Second, psychologists are too polite with one another's ideas. (Economists, for example, in my experience, don't frequently commit this particular sin.) In 2013, a prominent journal in psychology published a paper reporting the results of attempts to replicate a previously published finding. The title of the article was, before the colon, the phenomenon in question and then, after the colon: "Real or Elusive Phenomenon?" The pairing of "real" versus "elusive" (as opposed to "nonexistent") suggests that it's considered rude to label a result as a false positive rather than something that's simply hard to replicate—so rude that people in the field won't even say out loud that the prior work may have pointed to something that isn't, really, there.

Of course, intuitions interfere with theoretical innovation in other disciplines. No doubt the obviousness of the sun going around the Earth, bending across the sky each day, delayed acceptance of the heliocentric model. Everyone knows the mind isn't a hydraulic shovel, but it does feel as though some sort of reservoir of stuff gets used up, just as it does feel like the sun is moving and we're not.

Still, it's time that Cartesian hydraulicism be put to rest in the same way Cartesian dualism was.

THE COMPUTATIONAL METAPHOR

RODNEY A. BROOKS

Emeritus Panasonic Professor of Robotics, MIT; founder, chairman, and CTO, Rethink Robotics; author, Flesh and Machines

Throughout history, we have used technological systems as metaphors to describe how the body and brain might work. Ancient Greek water technology led to the four humors and the idea that they must be kept in balance. By the 18th century, clock mechanisms and flows of fluids were used as metaphors for what happened in the brain. By the first half of the 20th century, a common metaphor for the brain was a telephone switching network. Indeed, the mathematics developed for signal propagation in telegraph and telephone wires was used to model action potentials in axons. By the 1960s, cyberneticians were using models of negative feedback—originally developed for the steam engine and greatly expanded during the Second World War for controlling the aiming of guns—to try to develop models for the brain. But these soon ran out of steam, so to speak, and were supplanted by metaphors of the brain as a digital computer. We started to hear claims of the brain as the hardware and the mind as the software—a model that didn't much help our understanding of either brain or mind. Throughout the later 20th century, the brain became a massively parallel digital supercomputer, and now you can find claims that the brain and the World Wide Web are similar in how they work, with Web pages and neurons playing similar roles and hyperlinks and synapses mapping to each other.

This suggests that metaphors for the brain will continue to evolve as technology evolves, with the brain always corresponding to the most complex technology we currently possess. But does the metaphor of the day have an effect on the science of the day? I claim that it does, and that the computational metaphor leads researchers to ask questions today that will one day seem quaint, at best.

The power of computation and computational thinking is immense, and its import for science is still in its infancy. But it's not always helpful to confuse computational approximations with computational theories of a natural phenomenon. For instance, consider a classical model of a single planet orbiting a sun. There's a gravitational model, and the behavior of the two bodies can easily be explained as the solution to a simple differential equation describing forces and accelerations and their relationships. The equations can be extended for relativity and for multiple planets, and instantaneously those equations describe what a physicist would say is happening in the system. Unfortunately, the equations have become insoluble by this point, and the best we can do to understand the long-term behavior of the system is to use computation, where time is cut into slices and a digital approximation to the continuous description of the local behaviors is used to run a long-term simulation. However, only the most diehard of computationalists (and they do exist) would claim that the planets themselves are "computing" what to do at each instant. We know that it's more fruitful to continue to think of the planets as moving under the influence of gravity.

When it comes to explaining the brain and simpler neural systems, the computational metaphors have taken over. We hear people talk about neural coding. What is it that's coded in the spike train running along an axon over time? But early neurons evolved to synchronize muscle activity better. For instance, jellyfish swim much better if all their swimming muscle activates at once, so that they go straight rather than wobble, and evolution found multiple solutions in different species for this problem. The solutions range from really fast spike propagation to carefully tuned attenuation of signals along the triggering axon and local delays at the muscle fibers dependent on spike strength. Furthermore, in many jellyfish there are multiple neural systems based on different propagation chemistries for different behaviors and even for different modes of swimming. Just as describing planets as computational systems isn't the best way to understand what's going on, thinking of neurons in these simple systems as computational systems sending "messages" to one another isn't the best way to describe the behavior of the system in its environment.

The computational model of neurons of the last sixty-plus years excluded the need to understand the role of glial cells in the behavior of the brain, or the diffusion of small molecules affecting nearby neurons, or hormones as ways that different parts of neural systems affect each other, or the continuous generation of new neurons, or countless other things we haven't yet thought of. They didn't fit within the computational metaphor, so for many they might as well not exist. The new mechanisms we do discover outside of straight computational metaphors get pasted on to computational models, but it's becoming unwieldy—and worse, that unwieldiness is hard to see for those steeped in its traditions, racing along to make new publishable increments to our understanding. I suspect we'll be freer to make new discoveries when the computational metaphor is replaced by metaphors that help us understand the role of the brain as part of a behaving system in the world. I have no clue what those metaphors will be, but the history of science tells us they'll eventually come along.

LEFT-BRAIN/RIGHT-BRAIN

SARAH-JAYNE BLAKEMORE

Royal Society University Research Fellow, professor of cognitive neuroscience, University College London; coauthor (with Uta Frith), The Learning Brain

Most people will have heard about the left-brain/right-brain idea. Maybe they've been told they're too "left-brained" and they want to be more "right-brained." The idea has made it into everyday parlance, has infiltrated schools everywhere, sells a lot of self-help books, and has even been used as the basis of scientific theories—for example, regarding gender differences in the brain. Yet it's an idea that makes no physiological sense.

Scientific lingo about how the two sides of the brain—the hemispheres—function has permeated mainstream culture, but the research is often overinterpreted. The notion that the brain's two hemispheres are involved in different "modes of thinking" and that one hemisphere dominates the other has become widespread—in particular, in schools and the workplace. There are numerous websites where you can find out whether you're left-brained or right-brained and which offer to teach you how to change this.

This is pseudoscience and not based on knowledge of how the brain works. While it's true that the brain is made up of two hemispheres and one hemisphere is often initially active before the other during actions, speech, and perception, both sides of the brain work together in almost all situations, tasks, and processes. The hemispheres are in constant communication with each other, and it simply isn't possible for one hemisphere to function without the other joining in, except in certain rare patient populations. In other words, you're not right- or left-brained. You use both sides of your brain.

Some have proposed that education currently favors left-brain modes of thinking, which are supposed to be logical, analytical, and accurate, and doesn't put enough emphasis on right-brain modes of thinking, which are supposed to be creative, intuitive, emotional, and subjective. Certainly education should involve a wide variety of tasks, skills, learning, and modes of thinking. However, it's just a metaphor to refer to these as right-brain or left-brain modes. Patients who have had a lesion in their right hemisphere aren't devoid of creativity. Patients with a damaged left hemisphere might be unable to produce language (which relies on the left hemisphere in over 90 percent of the population) but can still be analytical.

Whether left-brain/right-brain notions should influence the way we educate is highly questionable. There's no validity in categorizing people, in terms of their abilities, as either left-brained or right-brained. Such categorization might even act as an impediment to learning—not least because it might be interpreted as innate, or fixed to a large degree. Yes, there are individual differences in cognitive strengths. but the idea that people are left-brained or right-brained needs to be retired.

LEFT-BRAIN/RIGHT-BRAIN

STEPHEN M. KOSSLYN

Psychologist; founding dean, Minerva Schools, Keck Graduate Institute; coauthor (with G. Wayne Miller), Top Brain, Bottom Brain

Solid science sometimes devolves into pseudoscience, but the imprimatur of being science nevertheless may remain. No better example of this is the popular "left brain/right brain" narrative about the specializations of the cerebral hemispheres. According to this narrative, the left hemisphere is logical, analytic, and linguistic whereas the right is intuitive, creative, and perceptual. Moreover, each of us purportedly relies primarily on one half-brain, making us "left-brain thinkers" or "right-brain thinkers." This characterization is misguided, and it's time to put it to rest.

Two major problems can be identified at the outset:

First, the idea that each of us relies primarily on one or the other hemisphere is not empirically justifiable. The evidence indicates that each of us uses the whole brain, not primarily one side or the other. The brain is a single, interactive system, whose parts work in concert to accomplish a given task.

Second, the functions of the two hemispheres have been mischaracterized. Without question, the two hemispheres engage in some different kinds of information processing. For example, the left preferentially processes details of objects we see, whereas the right preferentially processes the overall shape of objects we see. The left preferentially processes syntax (the literal meaning); the right preferentially processes pragmatics (the indirect or implied meaning). And so forth.

Our two hemispheres aren't like our two lungs: One isn't a "spare" for the other, redundant in function. But none of these well-documented hemispheric differences come close to what's described in the popular misconception.

It's time to move past the incorrect left brain/right brain narrative.

MOORE'S LAW

ANDRIAN KREYE

Editor, The Feuilleton *(arts and essays), of the German daily* Süddeutsche Zeitung, *Munich*

Gordon Moore's 1965 paper stating that the number of transistors on integrated circuits will double every two years has become the most popular scientific analogy of the digital age. Despite being a mere conjecture, it has become the go-to model to frame complex progress in a simple formula. There are good technological reasons to retire Moore's Law: for example, the consensus that it will effectively cease to exist past a transistor size smaller than 5 nanometers—which would mean a peak and sharp drop-off in ten to twenty years. Another is the potential of quantum computers to push computing into new realms; they're expected to become reality in three to five years. But Moore's Law should be retired *before* its technological limits, because it has sent the perception of progress in wrong directions. Treating its end as an event would just amplify the errors of reasoning.

First and foremost, Moore's Law has led us to see the development of the digital era as a linear narrative. The simple curve of progression is the digital equivalent of the ancient wheat-and-chessboard problem (with a potentially infinite chessboard). Like the Persian inventor of the game of chess who demanded from the king a geometric progression of grains all across the board, digital technology seems to develop exponentially. This model ignores the parallel nature of digital progress, which encompasses not only technological and economic development but scientific, social, and political change—changes that can rarely be quantified.

Still, the Moore's Law model of perception has already found its way into the narrative of biotechnological history, where change becomes ever more complex. Proof of progress is claimed in the simplistic reasoning of a sharp decline in cost for sequencing a human genome: from $3 billion in the year 2000 to the August 2013 cancellation of the Genomics X Prize for the first $1,000 genome because the challenge had been outpaced by innovation.

For both digital and biotechnical history, the linear narrative has been insufficient. The prowess of the integrated circuit has been the technological spark to induce a massive development comparable with the wheel allowing the rise of urban society. Both technologies have been perfected over time, but their technological refinement falls short when it comes to illustrating the impact both had.

Some twenty-five years ago, scientists at MIT's Media Lab told me about a paradigmatic change in computer technology. In the future, they said, the number of other computers connected to a computer will be more important than the number of transistors on its integrated circuits. To a writer interested in but not part of the forefront of computer technology, that was groundbreaking news. A few years later, the demonstration of a Mosaic browser was as formative for me as listening to the first Beatles record and seeing the first man on the moon had been for my parents.

Changes since have been so multilayered, interconnected, and rapid that comprehension has lagged behind. Scientific, social, and political changes occur in random patterns. Results have been mixed, in equally random patterns. The slowdown of the music industry and media has not been matched in the publishing and film industry. The failed Twitter revolution in Iran had much in common with the Arab Spring, but even in the Maghreb the results differed wildly. Social networks have affected societies sometimes in exactly opposite ways: Whereas the fad of social networks has resulted in cultural isolation in Western society, it has created a counterforce of collective communication against the strategies of the Chinese Party apparatus to isolate its citizenry from within.

Most of these phenomena have only been observed, not yet explained. It's mostly in hindsight that a linear narrative is constructed. The inability of many of the greatest digital innovations—like viral videos or social networks—to be monetized is just one of many proofs of how difficult it is to get a comprehensive grasp of digital history. Moore's Law and its numerous popular applications to other fields of progress thus create an illusion of predictability in the least predictable of all fields—the course of history.

These errors of reasoning will be amplified if Moore's Law is allowed to come to its natural end. Peak theories have become the lore of cultural pessimism. If Moore's Law is allowed to become a finite principle, digital progress will be perceived as a linear progression toward a peak and an end. Neither will become a reality, because the digital realm is not a finite resource but an infinite realm of mathematical possibilities reaching out into the analog world of sciences, society, economics, and politics. Because this progress has ceased to depend on a quantifiable basis and linear narratives, it won't be halted, not even slowed down, if one of its strains comes to an end.

In 1972, the wheat-and-chessboard problem became the mythological basis for the Club of Rome's Malthusian *Limits to Growth.* The peaking of Moore's Law will create the illusion that the digital realm could be a world of limits and finite resources. This doomsday scenario could become as popular as the illusion of predictable advances it created. After all, there have never been loonies carrying signs saying, "The End is Not Near."

THE CONTINUITY OF TIME

ERNST PÖPPEL

Psychologist, neuroscientist; cofounder, Human Science Center, Munich University; author, Mindworks

Philosophiae Naturalis Principia Mathematica, by Isaac Newton, is one of the fundamental works of modern science, and this is true not only for physics but also for philosophy and the foundations of reasoning. Newton gives, in "Scholium," the following definition: "Absolute, true, and mathematical time, of itself, and from its own nature, flows equably without relation to anything external." The underlying concept of continuity of time is expressed in mathematical formulas describing, for instance, physical processes. This concept of continuity is almost never questioned.

The Newtonian concept of the continuity of time is also implicitly assumed by Immanuel Kant when he refers to time as an "*a-priori* form of perception" in his *Critique of Pure Reason.* We read in the translation: "Time is not an empirical concept. For neither coexistence nor succession would be perceived by us, if the representation of time did not exist as a foundation *a priori*. . . . Time is a necessary representation, lying at the foundation of all our intuitions."

The concept of the continuity of time is also hidden in another famous quotation in psychology. William James writes, in *The Principles of Psychology:*

> In short, the practically cognized present is no knife-edge, but a saddle-back, with a certain breadth of its

> own on which we sit perched, and from which we look in two directions into time. The unit of composition of our perception of time is a *duration*, with a bow and a stern, as it were—a rearward- and a forward-looking end. . . . [W]e seem to feel the interval of time as a whole, with its two ends embedded in it.

Here we're confronted with the idea of a traveling moment: That is, a temporal interval of finite duration is moving gradually through physical time (and not jumping), again assuming continuity of time. But is it really true, and can it be used to understand neural and cognitive processes?

This theoretical concept of continuity of time in biological and psychological processes—usually appearing as an implicit assumption or "unasked question"—is wrong. The answer is simple if one takes a look at the way organisms process information to overcome the complexity and temporal uncertainty of stimuli in the physical world. One source of complexity comes from stimulus transduction, which is principally different in the sensory modalities like audition or vision, taking less than 1 millisecond in the auditory system and more than 20 milliseconds in the visual system. Thus, auditory and visual signals arrive at different times in central structures of the brain.

Matters become more complicated by the fact that the transduction time in the visual modality is flux-dependent, since surfaces with less flux require more transduction time at the receptor surface. Thus, to see an object with areas of different brightness, or to see somebody talking, different temporal availabilities of local activities within the visual modality and similarly different local activities across the two modalities engaged in stimulus processing must be overcome. For inter-sensory integration, aside from these biophysical problems, physical problems also have to be considered. The distance of objects to be perceived is obviously never predetermined. Thus, the speed of sound (not of light) becomes a critical factor.

At a distance of approximately 10 to 12 meters, transduction time in the retina, under optimal optical conditions, corresponds to the time the sound takes to arrive at the recipient. Up to this "horizon of simultaneity," auditory information is earlier than visual information; beyond this horizon, visual information arrives earlier in the brain. Again, there must be some kind of mechanism that overcomes the temporal uncertainty of information represented in the two sensory modalities. How can this problem be solved? The best way is for the brain to step out of the mode of continuous information processing.

The brain has indeed developed specific mechanisms to reduce complexity and temporal uncertainty by creating system states (possibly using neuronal oscillations) within which "Newtonian time" does not exist. Within such system states, temporally and spatially distributed information can be integrated, as experimental evidence shows. These states are "atemporal" because the before-and-after relationship of stimuli processed within them is not defined or definable. This biological trick implies that time does not flow continuously but jumps from one atemporal system state to the next.

THE INPUT-OUTPUT MODEL OF PERCEPTION AND ACTION

ANDY CLARK

Philosopher, Chair in Logic and Metaphysics, University of Edinburgh; author, Supersizing the Mind

It's time to retire the image of the mind as a kind of cognitive couch potato—a passive machine that spends its free time just sitting there waiting for an input to arrive to enliven its day. When an input arrives, this view suggests, the system swings briefly into action, processing the input and preparing some kind of output (the response, which might be a motor action or some kind of decision, categorization, or judgment). Output delivered, the cognitive couch potato in your head slumps back awaiting the next stimulation.

The true story looks to be almost the reverse. Naturally intelligent systems (humans, other animals) are not passively awaiting sensory stimulation. Instead, they are constantly active, trying to predict the streams of sensory stimulation before those arrive. When an "input" (itself a dodgy notion) occurs, our proactive cognitive systems have already been busy predicting its shape and implications. Systems like that are already (pretty much constantly) poised to act, and all they need to process are any sensed deviations from the predicted state.

Action itself needs to be reconceived. Action is not so much a response to an input ("input-output-stop") as a neat and efficient way of selecting the next "input," driving a rolling cycle. These hyperactive systems are constantly predicting their own upcoming states and moving about so as to bring some of those states into being. In this way, we bring forth the evolving streams of sensory information that keep us viable (fed, warm, and watered) and serve our increasingly recondite ends.

As ever-active prediction engines, these kinds of minds are not, fundamentally, in the business of solving puzzles given to them as inputs. Rather, they're in the business of keeping us one step ahead of the game, poised to act, and actively eliciting the sensory flows that keep us viable and fulfilled.

Just about every aspect of the passive input-output model is thus false. We aren't cognitive couch potatoes so much as pro-active predictavores, forever trying to stay one step ahead of the incoming waves of sensory stimulation. Keeping this in mind will help us design better experiments, build better robots, and appreciate the deep continuities binding life and mind.

KNOWING IS HALF THE BATTLE

LAURIE R. SANTOS

Associate professor of psychology; director, Comparative Cognition Laboratory, Yale University

TAMAR GENDLER

Vincent J. Scully Professor of Philosophy; professor of psychology and cognitive science; deputy provost for Humanities and Initiatives, Yale University

Children of the 1980s (like the younger of these coauthors) may fondly remember a TV cartoon called *G. I. Joe*, whose closing conceit—a cheesy public service announcement—remains a much-parodied YouTube sensation almost thirty years later. Following each of these moralizing pronouncements came the show's famous epithet: "Now you know. And knowing is half the battle."

While there may be some domains where knowing *is* half the battle, there are many more where it is not. Recent work in cognitive science has demonstrated that knowing is a shockingly tiny fraction of the battle for most real-world decisions. You may know that $19.99 is pretty much the same as $20.00, but the first still feels like a significantly better deal. You may know that a prisoner's guilt is independent of whether you're hungry or not, but she'll still seem like a better parole candidate when you've recently had a snack. You may know that a job applicant of African descent is as likely to be as qualified as one of European descent, but the negative aspects of the former's résumé will still stand out. And you may know that a piece of fudge shaped like dogshit will taste delicious, but you'll still be hesitant to eat it.

The lesson of much contemporary research in judgment and decision making is that knowledge—at least in the form of our consciously accessible representation of a situation—is rarely the central factor controlling our behavior. The real power of online behavioral control comes not from knowledge but from things like situation selection, habit formation, and emotion regulation. This is a lesson that therapy has taken to heart but one that "pure science" continues to neglect.

And so the idea cognitive science needs to retire is what we'll call the *G. I. Joe fallacy*: the idea that knowing is half the battle. It needs to be retired not just from our theories of how the mind works but also from our practices of trying to shape minds to work better.

You might consider this old news. After all, thinkers for the last 2,500 years have been pointing out that much of human action isn't under rational control. Don't we *know* by now that the G. I. Joe fallacy is just that—a fallacy?

Well, yeah, we *know*, but . . .

The irony is that knowing the G. I. Joe fallacy is a fallacy is—as the fallacy would predict—less than half the battle. As is knowing that people tend to experience $19.99 as a significantly lower price than $20.00. Even if you *know* about this left-digit anchoring effect, the first item will still feel like a better deal. Even if you *know* about ego-depletion effects, the prisoner you encounter after lunch will still seem like a better candidate for parole. Even if you *know* that implicit bias is likely to affect your assessment of a résumé's quality, you will still deem the African-American candidate less qualified than the Caucasian. And even if you *know* about Paul Rozin's disgust work, you will still hesitate to drink Dom Perignon out of a sterile toilet bowl.

Knowing isn't half the battle for most cognitive biases, including the G. I. Joe fallacy. Simply recognizing that the G. I. Joe fallacy exists is insufficient to avoid its grasp.

So now you know. And that's less than half the battle.

INFORMATION OVERLOAD

JAY ROSEN

Associate professor of journalism, New York University

We should retire the idea that goes by the name "information overload." It's no longer useful.

The Internet scholar Clay Shirky puts it well: "There's no such thing as information overload. There's only filter failure."[t] If your filters are bad, there's too much to attend to and never enough time. These aren't trends powered by technology, they're conditions of life.

Filters in a digital world work not by removing what needs removing but simply by not selecting for it. The unselected material is still there, ready to be let through by someone else's filter. Intelligent filters, which is what we need, come in three kinds:

- A smart person who takes in a lot and tells you what you need to know. The ancient term for this is "editor." The front page of the *New York Times* still works this way.
- An algorithm that sifts through the choices other smart people have made, ranks them, and presents you with the top results. That's how Google works—more or less.
- A machine-learning system that over time gets to know your interests and priorities and filters the world for you in a smarter and smarter way. Amazon uses systems like that.

Here's the best definition of information I know of: Information is a measure of uncertainty reduced. The definition is deceptively simple. In order to have information, you need two things: an uncertainty that matters (we're having a picnic tomorrow; will it rain?) and something that resolves it (the weather report.) But some information creates the uncertainty that needs solving.

Suppose we learn from news reports that the National Security Agency broke encryption on the Internet. That's information. It reduces uncertainty about how far the U.S. government was willing to go (all the way). But the same report increases uncertainty about whether there will continue to be a single Internet, setting us up for more information when that larger picture becomes clearer. So information is a measure of uncertainty reduced but also of uncertainty created. Which is probably what we mean when we say, "Well, that raises more questions than it answers."

Filter failure occurs not from too much information but from too much incoming stuff that neither reduces existing uncertainty nor raises questions that count. The likely answer is to combine the three types of filtering: smart people who do it for us, smart crowds and their choices, smart systems that learn by interacting with us as individuals. At this point, someone usually shouts out, "What about serendipity?" That's a fair question. We need filters that listen to our demands but also let through what we have no way of demanding because we don't know about it yet. Filters fail when they know us too well and when they don't know us well enough.

THE RATIONAL INDIVIDUAL

ALEX (SANDY) PENTLAND

Toshiba Professor of Media, Arts, and Sciences, MIT; director, MIT's Human Dynamics Laboratory and Media Lab Entrepreneurship Program; author, Social Physics: How Good Ideas Spread

Researchers argue about the extent to which people are rational, but the real problem with the concept of the rational individual is that our desires, preferences, and decisions are not primarily the result of individual thinking. Because economics and much of cognitive science take the unit of analysis to be an independent individual, they have difficulty accounting for such social phenomena as financial bubbles, political movements, panics, technology trends, or even the course of scientific progress.

Near the end of the 1700s, philosophers began to declare that humans were rational individuals. People were flattered by being recognized as individuals and by being called rational, and the idea soon wormed its way into the belief systems of nearly everyone in upper-class Western society. Despite resistance from church and state, this idea of rational individuality replaced the assumption that truth came only from god and king. Over time, the ideas of rationality and individualism changed the entire belief system of Western intellectual society, and today it's doing the same to the belief systems of other cultures.

Recent research data from my lab and other labs are changing this argument, and we're now coming to realize that human behavior is determined as much by social context as by rational thinking or individual desires. Rationality, as economists use the term, means that an individual knows what he or she wants and acts to get it. But this new research shows that in this regard, social-network effects often, and perhaps typically, dominate both the desires and the decisions of individuals.

Recently, economists have moved toward the idea of "bounded rationality," which means that we have biases and cognitive limitations that prevent us from realizing full rationality. Our dependence on social interactions, however, isn't simply a bias or a cognitive limitation. Social learning is an important method of enhancing individual decision making. Similarly, social influence is central to constructing the social norms enabling cooperative behavior. Our ability to survive and prosper is due to social learning and social influence at least as much as to individual rationality.

These data tell us that what we want and value, and how we choose to act in order to obtain our desires, are a constantly evolving property of interactions with other people. Our desires and preferences are mostly based on what our peer community agrees is valuable, rather than on rational reflection based directly on our individual biological drives or inborn morals.

For instance, after the Great Recession of 2008, when many houses were suddenly worth less than their mortgages, researchers found that it took only a few people walking away from their houses and mortgages to convince many of their neighbors to do the same. A behavior that had previously been thought of as nearly criminal or immoral—purposely defaulting on a mortgage—now became common. To use the terminology of economics: In most things we're collectively rational, and only in some areas are we individually rational.

By mathematically modeling the social learning and social

pressure among people, my colleagues and I have been able to accurately model and predict crowd phenomena such as this cascade of mortgage defaulting. Importantly, we've also found that it's possible to shape real-world crowd behaviors by using social-network incentives that alter the connections between people, and that these social incentives are much more effective than standard individual economic incentives. In one particularly striking example, we were able to use social-network incentives to deflate a "groupthink" bubble among foreign-exchange traders and consequently double the return on investment of the individual traders.

So instead of individual rationality, we have common sense. The collective intelligence of a community comes from the surrounding flow of ideas and examples; we learn from others in our environment, and those others learn from us. Over time, a community with members who actively engage with one another creates a group with shared integrated habits and beliefs. When the flow of ideas incorporates a constant stream of outside ideas as well, then the individuals in the community make better decisions than they could on their own.

This idea of a collective intelligence that develops within communities is an old one; indeed, it is embedded in the English language. Consider the word "kith," familiar to modern English speakers from the phrase "kith and kin." Derived from old English and old German words for knowledge, "kith" refers to a more-or-less cohesive group with common beliefs and customs. These are also the roots for "couth," which means possessing a high degree of sophistication, as well as its more familiar counterpart, "uncouth." Thus, our kith is the circle of peers (not just friends) from whom we learn the "correct" habits of action.

Our ancestors understood that our culture and the habits of our society are social contracts, and that both depend primarily on social learning. Thus we learn most of our public beliefs and habits by observing the attitudes, actions, and outcomes of peers rather than by using logic or argument. Learning and reenforcing this social contract enables a group of people to coordinate their actions effectively. It's time we dropped the fiction of individuals as the unit of rationality and recognized that we are embedded in the surrounding social fabric.

HOMO ECONOMICUS

MARGARET LEVI

Director, Center for Advanced Study in the Behavioral Sciences and professor of political science, Stanford University; coauthor (with John Ahlquist), In the Interest of Others

Homo economicus is an old idea and a wrong idea, deserving a burial of pomp and circumstance but a burial nonetheless. People can be individualistic and selfish, yes, and under some circumstances narrowly focused on economic well-being. But even those most closely associated with the concept never fully believed it. Hobbes argued that people prefer to act according to the Golden Rule but that their circumstances often made it difficult. Without rule of law and in a world of theft and predation, people act with defensive selfishness. Adam Smith, whose Invisible Hand required individual pursuit of narrow interest, recognized that individuals have emotions, sentiments, and morals that influence their thinking. Even Milton Friedman was not sure whether narrowly selfish individualism was a correct assumption about human behavior; he didn't care whether the supposition was right or wrong but only whether it was useful. It no longer is.

The theories and models derived from the assumption of *Homo economicus* generally depend on a second, equally problematic assumption: full rationality. Related but distinct sets of scientific findings make suspect each piece of the pairing of narrowly selfish motivation with rational action. Philosophers, such as Nietzsche, and psychoanalytic theorists, such as Sigmund Freud, argued that people acted in a whole variety of ways that were explicable perhaps but were closer to animal instincts than calculative instrumentality. Herbert Simon and certainly Daniel Kahneman and Amos Tversky revealed the extent to which cognitive limitations undermine rational calculations.

Even if individuals can do no better than "satisfice," that wonderful Simon term, they might still be narrowly self-interested, albeit (because of cognitive limitations) ineffective in achieving their ends. This perspective, which is at the heart of *Homo economicus*, must also be laid to rest. Darwin and those influenced by him long recognized that our species, like others, is altruistic at least in the narrow sense of acting to preserve one's gene pool by protecting one's young. Most people do much more than that. The overwhelming finding of experimental research confounds the presumption that given the opportunity individuals usually free-ride. Indeed, most act according to norms of fairness and reciprocity. Many will make small sacrifices or forgo larger returns, and some will even engage in costly action (up to a point) to "do the right thing." Anthropologists and biologists have long provided evidence of the human animal as a social animal. The understanding that individuals are in social networks and communities opens the door to more complex models of reciprocity and ethical obligation. Consequently, social scientists can now account for aggregate outcomes they otherwise could not: large-scale volunteering for the military in times of war, protest behavior, and contributions to public good.

The rejection of *Homo economicus* doesn't mean a total absence of conditions under which narrow self-interest dominates. Experiments suggest very different socializations can produce quite distinct reasoning: Economics graduate students are far more likely to free-ride than other students. At least two sets of circumstances can induce individualistic selfishness and significantly narrow a person's community of fate—that is, those with whom one feels interdependent and whom one feels an obligation to help. The first is extreme poverty and the second extreme competition. Those suffering hunger and deprivation tend to focus on meeting their own needs. As the growing number of dystopian novels suggests, the result may be theft and murder in the interest of obtaining food, shelter, and security. The classic experiments with rats come to the same conclusions.

Extreme competition, at the least, narrows focus to the goal at hand. In some forms, however, striving to be king of the hill, and sometimes literally to be king, does provoke something akin to a Hobbesian world. Shakespeare, as he often does, captures the power of circumstance and ambition; his version of the War of the Roses is a testament to narrow self-interested instrumentality dressed in the rhetoric of serving the country. Or witness the recent revelations about business ethics (or, rather, lack of ethics).

That people are often—perhaps more often than not—motivated to act beyond narrow self-interest is fully compatible with the importance of material incentives in motivating behavior. We're all susceptible to rewards, and we all fear punishment. *Ceteris paribus*, we prefer the first and wish to avoid the second. However, ethics, morality, and the obligations of reciprocity can affect our decisions even when there's considerable money at stake or serious threats to well-being. Few are willing to sacrifice everything for a cause or principle, but most of us are willing to sacrifice something.

The reliance on *Homo economicus* as the basis of human motivation has given rise to a grand body of theory and research over the past 200 years. As an underlying assumption, it has generated some of the best work in economics. As a foil, it has generated findings about cognitive limitations, the role of social interactions, and ethically based motivations. The power of the concept of *Homo economicus* was once great, but its power has now waned, succeeded by new and better paradigms and approaches grounded in more realistic and scientific understandings of the sources of human action.

DON'T DISCARD WRONG THEORIES, JUST DON'T TREAT THEM AS TRUE

RICHARD H. THALER

Charles R. Walgreen Distinguished Service Professor of Behavioral Science and Economics, Booth School of Business, University of Chicago; coauthor (with Cass R. Sunstein), Nudge: Improving Decisions About Health, Wealth, and Happiness

I have a problem with this year's *Edge* Question, so I'll answer a somewhat different question. I suppose the intent of the question is to point to some ideas that have been definitively shown to be either wrong or unhelpful and thus should be dropped from our scientific lexicon. In economics, there are certainly many theories, hypotheses, and models that are badly flawed descriptions of the behavior of economic agents, so you might think I'd have many nominations for ideas that should be given funerals. But I don't. Most of these theories, while demonstrably poor descriptions of reality, are extremely useful as theoretical baselines. As such, it would be a mistake to declare them dead.

Before getting to a couple of specific examples, I should stress that in economics, theories usually serve dual purposes. The first is "normative," in the sense that it defines what a rational agent *should* do. The second is "descriptive"; that is, it's meant to be an accurate description of how firms actually behave. Economists use the same theory for both purposes, and this leads to problems.

Consider the efficient-market hypothesis (EMH), first elaborated by my Chicago colleague Eugene Fama, who recently received the Nobel Prize in economics. The theory has two components. The first is that prices are unpredictable and you can't beat the market. I call this the No Free Lunch part of the EMH. The second is that asset prices are equal to fundamental value. I call this the Price is Right component. Ever since the EMH was formulated, it has been used as a baseline, null hypothesis in financial economics research. In a world consisting only of rational investors, both components would be descriptively accurate, but of course we don't live in such a world. How does the theory stand up in the real world?

If I were fact-checking the No Free Lunch part of the theory, I'd score it "mostly true." It *is* hard to beat the market, and most people who try fail, including professional mutual-fund managers. The No Free Lunch component is only "mostly" true because sometimes you *can* beat the market—for example, by buying "value stocks" whose prices seem low relative to earnings or assets. Still, a strategy of buying cheap index funds that track the market is a sensible one for investors to follow, so believing this part of the theory does little damage.

The second component of the theory, The Price is Right, is both more important and more problematic. Two recent experiences, the tech-stock bubble in the late 1990s and the real-estate bubble in the early 2000s, reveal that prices can diverge to a significant degree from their intrinsic value. The late financial economist Fischer Black, coinventor of the famous Black-Scholes option-pricing formula, once conjectured that asset prices can diverge from their true values by a factor of 2. Fischer, who died in 1995, might have revised that estimate to a factor of 3 had he lived to see the NASDAQ fall from 5,000 to 1,400 when the tech bubble burst. More than a decade later, the NASDAQ is only now reaching the 4,000 level, with no adjustment for inflation.

With the two components of the EMH graded partly wrong and badly wrong should we abandon the theory? Hardly. None of the research done by behavioral finance researchers—including my fellow traveler Robert Shiller, who shared the 2013 Nobel Prize with Fama along with Lars Hansen—would have been possible without the EMH benchmark. Shiller's early research showed that prices were too variable, compared to what would be expected in a rational model.

So, if we shouldn't banish the EMH, what should change? The change I'd advocate is abolishing the presumption that it's true. Part of Alan Greenspan's reasoning—after hearing a talk from Shiller in 1996 warning of an overheated market—for the Fed not taking any action was that bubbles were impossible in an efficient market. Even the Supreme Court, in the 1988 case *Basic, Inc., vs. Levinson,* ruled that plaintiffs could rely on the efficient-market hypothesis in bringing cases alleging misconduct by firms.

The problem here is that users of this concept are neglecting the last word in the phrase "efficient-market hypothesis." The same mistake is made in the use of another theory that contributed to a Nobel Prize: Franco Modigliani's life-cycle hypothesis. Here the hypothesis is that people figure out how much they're going to make over their lifetime, how much they'll earn on their investments, how long they'll live, and then solve for the optimal amount to save each year while they're accumulating money—and, similarly, at what rate to draw down their assets once they retire. Once again, this is a useful benchmark and can be helpful in offering advice to people regarding how much they should be saving for retirement.

It would be a mistake to discard this theory, but it would be a much bigger mistake to presume it's true. The hypothesis counterfactually assumes that people are capable of solving a very difficult mathematical problem and also of implementing such a plan without falling victim to spending temptations along the way. Presuming the theory to be true induced many economists to confidently, but wrongly, predict that offering people retirement savings plans such as 401(k)s would have no effect on savings, since people were already saving the right amount, and would merely shift their saving into the new tax-favored plans, costing the government money but producing no new saving. A similar presumption makes the false prediction that small changes such as automatically enrolling participants will have no effect on behavior.

Let's keep these and many other wrong theories and hypotheses alive, but remember they're just hypotheses, not facts.

RATIONAL ACTOR MODELS: THE COMPETENCE COROLLARY

SUSAN FISKE

Eugene Higgins Professor, Psychology and Public Affairs, Princeton University

The idea that people operate mainly in the service of narrow self-interest is already moribund, as social psychology and behavioral economics have shown. We now know that people aren't rational actors, instead often operating on automatic, based on bias, or happy with hunches. Still, it's not enough to make us into robots or to accept that we're flawed. The rational-actor corollary—that all we need is to show more competence—should also be laid to rest. Even regular people who aren't classical economists sometimes think that sheer cut-throat competence would be enough—on the job, in the marketplace, in school, and even at home.

Talent and problem-solving ability are crucial, of course. But there's more. We're social beings, embedded in a human environment even more than in a natural or a constructed one. If other people are our ecological niche, then we need to understand how to live among them. We do this by figuring out two things about them: not only "How good will they be at getting where they want to go?" but also "Where are they trying to go?"

People are a miracle of self-propelled agency. Not for nothing are humans attuned to one another's intentions. We need—and our ancestors needed—to know whether others have friendly or hostile intentions toward us. In my world, we call this a person's warmth, and others have called it trustworthiness, morality, communality, or worthy intentions.

People are most effective in social life if they are—and show themselves to be—both warm and competent. This is not to say we always get it right, but the intent and the effort must be there. This is also not to say love is enough, because we do have to prove able to act on our worthy intentions. The warmth-competence combination supports both short-term cooperation and long-term loyalty. In the end, it's time to recognize that people survive and thrive with *both* heart and mind.

MALTHUSIANISM

MATT RIDLEY

Science writer; founding chairman, International Centre for Life; author, The Rational Optimist: How Prosperity Evolves

T. Robert Malthus (he used his middle name) thought population must outstrip food supply unless checked by famine, disease, and war. So he warned that if people did not delay marriage, it would be necessary to "court the return of the plague" and "particularly encourage settlements in all marshy and unwholesome situations."[u] Alas, many people were only too happy to take up this nasty idea—that you had to be cruel to be kind to prevent population growing too fast for food supply. It directly influenced heartless policy in colonial Ireland, British India, imperial Germany, eugenic California, Nazi Europe, Lyndon Johnson's aid to India, and Deng Xiaoping's China. It was encountering the Club of Rome's Malthusian tract *The Limits to Growth* that led Song Jian to recommend a one-child policy to Deng. The Malthusian misanthropic itch is still around and far too common in science.

Yet Malthus' followers were wrong, wrong, wrong: Not just because they were unlucky that the world turned out nicer than they thought, not just because keeping babies alive proved a better way of getting birthrates down than encouraging them to die, not just because technology came to the rescue—but because Malthusians have repeatedly made the mistake of thinking of resources as static, finite things that would "run out." They thought growth meant using up a fixed heap of land, metals, water, nitrogen, phosphate, oil, and so forth.

They thought the birth of a calf was a good thing because it added to the world's resources, but the birth of a baby was a bad thing because it added to the mouths to feed.

They completely misunderstood the nature of a resource, which only becomes a resource thanks to human ingenuity. So uranium oxide was not a resource before nuclear power. Shale oil was not a resource until horizontal fracking. Steel was not easily recyclable until the electric-arc furnace. Nitrogen in the air was not a resource until the Haber process. The productivity of land was transformed by fertilizer, so globally we now use 65 percent less land than we did fifty years ago to produce the same amount of food. And a baby is a resource, too—a brain as well as a mouth.

The few economists, such as Julian Simon and Bjørn Lomborg, who tried to point this out to the Malthusian scientists—and who argued that economic growth was not the cumulative use of resources but the increase of productivity, doing more with less—were called imbeciles or had pies thrown in their face for their trouble. But they were right again and again, as population and prosperity grew together to levels that the Malthusians kept saying were impossible. "It is unrealistic to suppose that there will be increases in agricultural production adequate to meet forecast demands for food," said a long list of scientific stars in a British book called *A Blueprint for Survival* in 1972. "Farmers can no longer keep up with rising demand; thus the outlook is for chronic scarceties and rising prices" and famine is inevitable, said Lester Brown in 1974. (World food production has since doubled, and famine is largely history except where dictators create it).

World population will almost certainly cease to grow before the end of the century. Peak farmland is very close, if not already past—that is to say, we're using less farmland to grow more food, so we'll need less land, not more. Electric cars driven by nuclear power stations are, to all intents and purposes, an infinite resource. The world is a dynamic, reflexive place in which change is all. Time to retire the static, myopic, misanthropic mistakes done in the name of the mathematical Parson Malthus.

ECONOMIC GROWTH

CESAR HIDALGO

Assistant professor, MIT Media Lab; faculty associate, Center for International Development, Harvard University

Economic growth is one of those concepts nobody wants to contradict. Even its detractors cannot avoid using it. They talk about green growth, sustainable growth, and in the most extreme cases they talk about de-growth.

Yet economic growth, as a concept and a reality, is recent. Modern measures of economic growth are less than a century old, dating back to the invention of GDP by Simon Kuznets in the 1930s. Economists mostly agree that economies did not grow before the 19th century, so economic growth as a phenomenon is also recent.

As do many others, I believe that the idea of economic growth is ready for retirement. The lingering question is what will replace it, since economic growth will leave a void in public speech, as both a staple paragraph of political campaigns and a recurrent topic in news media. But economic growth cannot last forever. If the GDP per capita of the United States grew in real terms at a modest rate of 1 percent for the next millennium, the average American would be making a whopping $1.1 billion annually by the year 3014. A more reasonable interpretation of this number is to think of the growth during the last century as part of an S-shaped curve, a phase transition. This means either that growth will peter out during this millennium or that we're measuring the wrong thing. Either way, we can conclude that the idea of economic growth is on its way out.

UNLIMITED AND ETERNAL GROWTH

HANS ULRICH OBRIST

Curator, Serpentine Gallery, London; coauthor (with Rem Koolhaas): Project Japan: Metabolism Talks*; editor,* Do It: The Compendium

While studying political economy during the late 1980s, I was deeply inspired by the pioneer of ecology and economics Hans Christoph Binswanger, who is now in his eighties and is being rediscovered by younger artists and activists (e.g., Tino Sehgal), who often cite him as an influence.

The wisdom of Bingswanger's work is that he recognized early that endless growth is unsustainable, both in human and planetary terms. The current focus in mainstream economics is, he argues, too much on labor and productivity and too little on natural and intellectual resources. Dependency on endless growth—as the crisis that always emerges at the end of each cyclical bull market should teach us—is unrealistic.

Binswanger's goal was to investigate the similarities and differences between aesthetic and economic values through an examination of the historical relationship between economics and alchemy, which he made as interesting as it sounds (at first) outlandish. In his 1985 book *Money and Magic*, he showed how the brash concept of unlimited growth was inherited from the medieval discourse of alchemy, the search for a process that could turn lead into gold.

A focus of Binswanger's research has been on Goethe, especially his role in shaping social economics while finance minister at the court of Weimar. In Goethe's *Faust*, the eponymous character thinks in terms of infinite progress, while Mephisto recognizes the destructive potential of such an idea. At the beginning of part two of the play, Mephistopheles urges the ruler of an empire facing financial ruin because of profligate government spending to issue promissory notes, thus solving its debt problems. Binswanger had been fascinated by the Faust legend since his childhood, and during his studies he discovered that Goethe's introduction of paper money into his play was inspired by the story of the Scottish economist John Law, who in 1716 was the first man to establish a French bank issuing paper money. Strikingly, after Law's innovation, the Duke of Orleans got rid of all his alchemists, because he realized that the immediate availability of paper money was far more powerful than any attempt to turn lead into gold.

Binswanger also connects money and art in a novel way. Art, he points out, is based on imagination and is part of the economy, while a bank's process of creating money in the form of promissory notes or coins is connected to imagination, since it's based on the idea of bringing into being something that has yet to exist. At the same time, a company imagines producing a certain good and needs money to realize this, so it takes out a loan from a bank. If the product is sold, the "imaginary" money that was created in the beginning has a countervalue in real products.

In classical economic theory, this process can be continued endlessly. Binswanger recognizes, in *Money and Magic*, that this endless growth exerts a quasi-magical fascination. He produces a way of thinking about the problems of rampant capitalist growth, encouraging us to question the mainstream theory of economics and to recognize how it differs from the real economy. But instead of rejecting the market wholesale, he suggests ways in which to moderate its demands. Thus the market doesn't have to disappear or be replaced but can be understood as something to be manipulated for human purposes rather than obeyed.

Another way of interpreting Binswanger's ideas is as follows: For most of human history, a fundamental problem has been the scarcity of material goods and resources, and so we have become ever more efficient in our methods of production and created rituals to enshrine the importance of objects in our culture. Less than a century ago, human beings made a world-changing transition through their rapacious industry. We now inhabit a world in which the overproduction of goods, rather than their scarcity, is one of our most fundamental problems. Yet our economy functions by inciting us to produce more and more with each passing year. In turn, we require cultural forms to enable us to sort through the glut, and our rituals are once again directed toward the immaterial, toward quality and not quantity. This requires a shift in our values, from producing objects to selecting among those that already exist.

THE TRAGEDY OF THE COMMONS

LUCA DE BIASE
Journalist; editor, Nova 24, Il Sole 24 Ore

The tragedy of the commons is at an end, thanks to the writings of the late Nobel laureate Elinor Ostrom. But the well-deserved funeral has not been celebrated, yet. Thus, some consequences of the now-disproved theory proposed by Garrett Hardin in his famous 1968 article are still to be fully digested. Which is urgent, because some major problems we face in our age are very much related to the commons: climate change, the issue of privacy and freedom on the Internet, the choice between copyright or public domain in scientific knowledge.

Of course the commons can be overexploited. But what's wrong with Hardin's theory is the notion of "tragedy": By using that term, Hardin implied that a sort of destiny condemned the commons to be depleted. In his opinion, a big enough set of rational individuals who are free to choose will act in a way that will inevitably lead to the exhaustion of the commons, because free and rational individuals will always maximize their private advantage and collectivize the costs. Ostrom has demonstrated that this tragic destiny doesn't need to be true. She found, all over the world, an impressive number of cases in which communities run the commons in a sustainable way, getting the most out of them without depleting them.

Ostrom's factual approach to the commons came with very good theory, too. Preconditions to the commons' sustainability were, in Ostrom's view, clarity of the law, methods of collective and democratic decision making, local and public mechanisms of conflict resolution, and no conflicts among the various layers of government. These preconditions exist in many situations, and there is no tragedy there. Cultures that understand the commons are contexts that make a sustainable behavior rational.

Hardin's approach, which he developed during the cold war, was probably biased by ideological dualism. The commons didn't fit, as Ostrom noted in her Nobel lecture, in a "dichotomous world of 'the market' and 'the state'." In a context in which private property and deregulation versus state-owned resources and regulation were seen as the only two possible solutions, the commons was considered a losing system, condemned to the past.

But the Internet has become the biggest commons of knowledge in history. It would be difficult to argue that the Internet is a losing system. In the last twenty years, the commons of the Internet has changed the world. Of course, the Internet can be exploited by gigantic companies or state-owned secret services. But no tragic destiny condemns the Internet to ruin. To save it, we can start by understanding and preserving its rules, such as net neutrality, multi-stakeholder governance, and transparent enforcement of those rules and governance. Wikipedia has demonstrated that this is possible.

There's no tragedy, but there are conflicts. And they can be better understood by embracing a vision open to Ostrom's notion of polycentric governance of complex economic systems. The danger of a closed vision—one that understands only the conflicts between state regulation and market freedom—seems even more catastrophic when you contemplate climate change and other environmental issues. With regard to the environment, the commons idea seems a much more generative notion than many others. It doesn't guarantee a solution, but it's a better starting point. The "tragedy of the commons" theory has now become a comedy. But it can be a sad comedy if we don't finish with it and move on.

MARKETS ARE BAD; MARKETS ARE GOOD

MICHAEL I. NORTON

Associate professor of business administration, Marketing Unit, Harvard Business School; coauthor (with Elizabeth Dunn), Happy Money

Markets can have terrible consequences. Take just one example. In an ingenious experiment, researchers showed that people who entered a market where livestock were priced as commodities were more likely to devalue the lives of those animals—treating those lives as nothing more than opportunities for profit.

Markets can have uplifting consequences. Take just one example. In a series of investigations, researchers showed that efficient markets have contributed to the development of countless life-saving drugs (albeit sometimes with a little governmental help), bettering the lives of billions.

Yet in popular and scientific discourse, it's uncommon to see markets described as anything except evil and fundamentally flawed (left-leaning pundits and scholars) or perfect and self-correcting (right-leaning pundits and scholars). It's time to retire both theories.

Taking a step back and seeing markets as they are—an aggregation of many individuals—shows them as unlikely to be good or bad. Replace the word "markets" with another shorthand term for an aggregation of individuals: "groups." We don't view groups as good or bad. Groups are capable of amazing selflessness, generosity, and heroism; they're also capable of selfishness, greed, and cruelty. They're capable of amazing performance (think of Bell Labs); they're also capable of terrible performance (think of the many dysfunctional groups of which you've been a member).

When we think of groups, we think of the conditions under which they're likely to behave well or poorly. We don't often think of them as self-correcting, or always performing well over time, or (most important) as either inherently good or inherently bad. Applying the same logic to markets will help us develop a richer and more accurate theory of when and why they're likely to have terrible or uplifting consequences.

STATIONARITY

GIULIO BOCCALETTI

Physicist, atmospheric and oceanic scientist; managing director, The Nature Conservancy

When the ancient capital of the Nabataeans, Petra, was "rediscovered" by Johann Burckhardt in the early 1800s, it might have seemed unthinkable that anybody could have lived in such an arid place. Yet at its peak in the 1st century B.C.E., Petra was the center of a powerful trading empire and home to up to 30,000 people.

Petra's very existence was a testament to how water management could support the development of civilization in extreme circumstances. This part of the world—today in the Hashemite Kingdom of Jordan—survives on less than 70 mm of rain a year, much of it concentrated in a few events in the rainy season. The climate 2,000 years ago was similar, yet Petra thrived, thanks to a system of rock-cut underground cisterns, terraced slopes, dams, and aqueducts, which stored and delivered water from springs and runoff flows. Petra could grow food, provide drinking water, and support a bustling city because of that infrastructure.

This story is not dissimilar from that of many other places across the world today. The western United States, northern China, South Africa, the Punjab—such areas have thrived and grown, thanks to human ingenuity and water engineering, allowing people to overcome the adversity of a difficult, at times impossible, hydrology.

Whether the Nabataean engineers knew it or not, to

deliver reliable water infrastructure they relied—like all water engineers since—on two commonly assumed properties of hydrological events: stationarity and, rather more esoterically, ergodicity. Both concepts have well-defined mathematical meaning. Simply put though, stationarity implies that the probability distribution of a random event is independent of time. And a stationary process is ergodic if, given a sufficiently long time, it will realize most of the universe of options available to it.

Practically, this lets us assume that if we have observed an event for long enough, we'll also probably have witnessed enough of its behavior to represent the underlying distribution function at any given point in time. In the case of hydrology, stationarity lets us define events by using time statistics like the "one in a hundred years" flood.

The assumption that hydrology can be represented by such stationary processes enables the design of infrastructure whose behavior can be known well into the future. After all, such water infrastructure as dams, levees, and so on last for decades, even centuries, so it's important that they be constructed to withstand most predictable events. This is what allowed Nabataean, Chinese, American, South African, and Indian water engineers to design water systems they could legitimately rely on. And those systems have been successful—so far.

Stationarity provides a convenient simplifying gambit: that plans for future water management can be based on an appropriately long historical time series of past hydrology, because the past is simply a representative sequence of realizations of a (roughly) fixed probability distribution. But in the real world—where there's no counterfactual and a single experiment is running all the time—such assumptions are true only until proved wrong. We're now realizing that they *are* in fact wrong. Not just theoretically wrong but practically flawed.

A growing number of recent observations support the idea that probability distributions we assumed were fixed are not. They're changing, and changing fast. Many of what used to be one-in-a-hundred-year events are more likely to be one-in-twenty-year events. Droughts once considered extreme and unlikely are now much more common. Accelerating changes in climate, coupled with a more sensitive global economy, in which more people and more value is at stake, reveal that we don't live in a world as stationary as we thought. And infrastructure designed for that world and intended to last for decades is proving increasingly inadequate.

The implications are monumental for our relationship with the planet and its water resources. A broadly stationary environment can be "engineered away"; someone will take care of it as long as we can define what we need and have enough resources to pay for it. Things are different in a nonstationary world. The problem of water management is no longer decoupled from the dynamics of climate, as the climate is no longer constant on practical time scales. We face unforeseen variability. The past is no longer necessarily a guide to the future, and we cannot simply rely on "someone taking care of it." "It" is no longer just an engineering problem. Climatology, hydrology, ecology, and engineering all become relevant instruments in the management of a dynamic problem whose nature requires adaptability and resilience. Our own economy should be prepared to adapt, because no long-term piece of infrastructure can be expected to manage what it was not designed for.

By the 1st century C.E., the Nabataeans had been incorporated into the Roman Empire, and over subsequent centuries their civilization withered away, the victim of changing trade routes and shifting geopolitics (and proof that while water can support the development of civilizations, it's far from sufficient to see them thrive!). Today we have hundreds of cities around the world that, just like Petra, rely on engineered water infrastructure to support their growth. From Los Angeles to Beijing, from Phoenix to Istanbul, great cities of the world depend on a reliable source of water in the face of unreliable hydrology.

If stationarity is indeed a thing of the past, water management is no longer a "white coats" business, something that can be taken care of in the background. We must consider choices, have contingency plans for events we might not have experienced, and realize that we might get it wrong. We must go from managing water to managing risk.

STATIONARITY

LAURENCE C. SMITH

Professor and Chair of Geography; professor of Earth and space sciences, UCLA; author, The World in 2050: Four Forces Shaping Civilization's Northern Future

Stationarity—the assumption that natural-world phenomena fluctuate with a fixed envelope of statistical uncertainty that doesn't change over time—is a widely applied scientific concept that's ready to be retired.

It had a good run. For more than a century, stationarity has been used to inform countless decisions aimed at the public good. It guides the planning and building codes for places susceptible to wildfires, floods, earthquakes, and hurricanes. It's used to determine how and where homes may be built, the structural strength of bridges, and how much premium people should pay for their homeowner's insurance policies. Crop yields are forecasted and, in the developed world, insured against catastrophic failures. And as more weather stations and river-level gauges are built and accumulate ever longer data records, our abilities to make such calculations get better. This saves lives and a great deal of money.

But a growing body of research shows that stationarity is often the exception, not the norm. As new satellite technologies scan the Earth, as more geological records are drilled, and as the instrument records lengthen, they commonly reveal patterns and structures inconsistent with a fixed envelope of random noise. Instead, there are transitions to various quasi-stable states, each characterized by a different set of physical conditions and associated statistical properties. In climate science, for example, we have discovered multidecadal patterns, like the Pacific Decadal Oscillation, an El Niño–like phenomenon in the North Pacific triggering far-reaching changes in climate averages that persist for decades (in the 20th century, for example, the PDO experienced "warm" phases in 1922–1946 and 1977–1998 and a "cool" phase in 1947–1976) with significant effects on water resources and fisheries. And anthropogenic climate change, induced by our steady ramping-up of greenhouse-gas concentrations in the atmosphere, is by definition the opposite of a fixed, stationary process. This imperils the basis of many societal-risk calculations, because as the statistical probabilities of the past break down, we enter a world that operates outside of expected and understood norms.

This recognition is not new among scientists but has been surprisingly slow to penetrate the practical world. For example, even as awareness and acceptance of climate change has grown, stationarity continues to serve as the default assumption in water-resource risk assessment and planning. Floodplain zoning continues to be designed around stationary concepts like the 100- and 500-year flood, despite known impacts of land-use conversion and urbanization on water runoff and the anticipated impacts of anthropogenic climate change. The civil engineering profession and most regulatory agencies around the world have been slow to acknowledge these changes and seek new approaches to address them. But viable alternatives exist—for example, using the precautionary, no-regrets "probable maximum flood" (PMF) method to design dams and bridges, and incorporation of more flexible "subjectivist Bayesian" probabilities in societal-risk calculations.

We can do better. Stationarity is dead, especially for our understanding of the world's water, food security, and climate.

THE CARBON FOOTPRINT

DANIEL GOLEMAN

Psychologist; author, Focus: The Hidden Driver of Excellence

Buy potato chips in London and a number on the bag will tell you that its carbon footprint equals 75 grams of carbon emissions. That label serves two excellent functions: It renders transparent the ecological impact of those chips and lowers the cognitive cost to zero of learning that impact. Such carbon-footprint ratings, in theory, allow shoppers to favor products with better impacts and companies to do the same with their operations. Well and good. Except that the footprint concept, intended to mobilize the mass changes we need, ignores fundamentals of human motivation, tending to stifle change, not encourage it.

It's time we moved beyond talking about carbon footprints, replacing the concept with a more precise measure of all the negative effects of a given human activity on planetary systems for sustaining life. And while we're at it, let's go easy on the very idea of any kind of "footprint"—the numbers are demoralizing. There's a more motivating replacement waiting in the wings: handprint.

First, the expanded footprint. While the dialog on global warming and its remedies focuses tightly on the carbon impact of our activities and energy systems—as measured by their carbon footprint—this focus skews the conversation. Technically, a carbon footprint represents the total global-warming effect of greenhouse-gas emissions from a given activity, system, or product. While carbon dioxide is the poster child for greenhouse gases, other such gases include methane, nitrous oxide, and ozone (not to mention vaporized water or the condensed form, clouds). To create a standardized unit for greenhouse-gas effects, all these varieties of emissions are converted into a carbon-dioxide equivalent.

Reasonable, but this doesn't go far enough: Why stop with carbon? There are several planet-wide systems that maintain life; climate change is but one of myriad ways that human activity harms the planet. There's ecosystem destruction; dead lakes and ocean from acidification; loss of biodiversity; nitrogen and phosphorous cycles; dangers from particulate load in air, water, and soil; pollution from man-made chemicals; and more.

All these problems arise because virtually all human systems for energy, transportation, construction, industry, and commerce are built on platforms that degrade those global systems. Calculating the overall ecological footprint of a given activity offers a more fine-tuned metric for the rate at which we're depleting all the global systems sustaining life on the planet, not just the carbon cycle.

Such metrics emerged from the relatively new science of industrial ecology, an amalgam of hard sciences like physics, chemistry, and biology with practical applications like industrial engineering and industrial design. This eco-math helps us perceive effects we're otherwise oblivious to. For instance, when industrial ecologists measure how much of the carbon footprint you remediate when you recycle the plastic yogurt container, the result is about 5 percent of the yogurt's carbon footprint. Most of the yogurt's carbon footprint results from the methane emitted by digesting cattle, not from the plastic container.

Second, there's the motivational problem. Evolution shaped the human brain to help our ancestors survive in an era when the salient threats were predators. Our perceptual system was not tuned to the macro- and microchanges that signal threats to the planetary support system. When it comes to these threats, we suffer from system blindness. While footprints offer a cognitive workaround that can help us make decisions favoring the planet, they too often have an unfortunate psychological effect: Knowing the planetary damage we do can be depressing and demotivating. Negative messages like this, research from fields like public health finds, lead many or most people to tune out. Better to give us something positive we can do than to shame or scare us.

Enter the "handprint," the sum total of all the ways we lower our footprint. To calculate a handprint, take the footprint as the baseline and then go a step further: Assess the amount ameliorated by the good things we do—recycle, reuse, bike instead of drive. Convince other people to do likewise. Or invent a replacement for a high-footprint technology, like the Styrofoam subsititute made from rice hulls and mycelium rather than petroleum.

The handprint calculation applies the same methodology as for footprints but reframes the total as a positive value: Keep growing your handprint and you're steadily reducing your negative effect on the planet. Make your handprint bigger than your footprint and you're sustaining the planet, not damaging it. Such a positive spin, motivational research tells us, will be more likely to keep people moving toward the target.

UNBRIDLED SCIENTIFIC AND TECHNOLOGICAL OPTIMISM

STUART PIMM

Doris Duke Chair of Conservation Ecology, Duke University; author, A Scientist Audits the Earth

Science and technology have made such spectacular improvements to our lives that it seems churlish to whinge about them. I understand the benefits better than most. My fieldwork is where "the other half" lives—the majority of the world's population too poor to have access to safe drinking water, antibiotics, and much (if any) electricity. I can go home, flip a switch, turn on the tap, and carry Cipro wherever I go. Just as natural selection picks past winners but brutally trims most mutations, so the science we love does not make every scientist in a white lab coat a hero. Many proposed scientific advances are narrow in their benefits, poorly thought out in the long term, and attention-getting or venally self-serving. Worst of all, optimism creates a moral hazard. When science promises it can fix everything, why worry about breaking things?

For example, discussions about fracking, and the supplies of cheap fossil fuels it may give us, pit the local, near-term threats of a new technology against obvious benefits. For the United States, the energy is here and not in some politically sketchy country that requires vast military adventures to defend. Or to invade—for surely we would not have invaded Iraq if its principal export had been cantaloupes.

So bravo for fracking? Hardly! Suppose this, or any fossil fuel, were cheap and environmentally entirely free of local concerns. It would further accelerate global carbon emissions and their increasingly serious consequences. Perversely, the better (cleaner, cheaper, faster) the technology, the worse the eventual problem of too much atmospheric carbon dioxide. Surely decades of cheap gas give us breathing space to develop and transition to sustainable energies? That's a gamble with disastrous consequences to our planet if we fail.

Won't new technologies soak up the carbon for us, allowing fossil fuels free rein? Only in the minds of those who seek huge research funds to pursue their ideas. The best and cheapest technology is what we ecologists call trees. Burning them contributes about 15 percent of the global carbon emissions, so reducing those—as Brazil has done so successfully in recent years—is altogether a good idea. Restoring deforested areas is also prudent and economical. Trees have been around since the Devonian.

Of the many dire effects of a much hotter planet, the irreversible losses are to biodiversity. Species extinction rates already run 1,000 times higher than normal. Climate disruption will inflate them further.

Optimists have the answer! The purest hubris is to raise the dead. "De-extinction" seeks to resurrect individual extinct species, usually charismatic ones. You know the plot. In the movie *Jurassic Park*, a tree extinct for millions of years delights the paleobotanist. Then a sauropod eats its leaves. We then learn how to re-create the animal itself. The movie is curiously silent on how to grow the tree, which at that size would be perhaps 100 or more years old, metaphorically overnight. To sustain a single sauropod, one would need thousands of trees of many species, as well as their pollinators and perhaps their essential symbiotic fungi.

Millions of species risk extinction. De-extinction can be only an infinitesimal part of solving the crisis that now sees species of animals (some large, but most tiny), plants, fungi, and microbes going extinct at 1,000 times their natural rates.

Proponents of de-extinction claim they want only to resurrect passenger pigeons and the Pyrenean ibex, not dinosaurs. They assume that the plants on which these animals depended survive so there's no need to resurrect them as well. Indeed, botanic gardens worldwide have living collections of an impressively large fraction of the world's plants—some extinct in the wild, others soon to be so. Their absence from the wild is thus more easily fixed than the absence of animals for which optimists tout de-extinction.

Perhaps so, but other practical problems abound: A resurrected Pyrenean ibex will need a safe home, not just its food plants. For those of us who attempt to reintroduce zoo-bred species that have gone extinct in the wild, one question tops the list: Where do we put them? Hunters ate this wild goat to extinction. Reintroduce a resurrected ibex to where it belongs and it will quickly become the most expensive cabrito ever eaten.

De-extinction is much worse than a waste: It sets up the expectation that biotechnology can repair the damage we're doing to the planet's biodiversity. Fantasies of reclaiming extinct species are always seductive. "Real" scientists, wearing white lab coats, use fancy machines with knobs and digital readouts to save the planet from humanity's excesses. There's none of the messy interactions with people, politics, and economics that characterize my world. There's nothing involving the real-world realities of habitat destruction, of the inherent conflict between increasing human populations and wildlife survival. Why worry about endangered species? We can simply keep their DNA and put them back in the wild later.

When I testify before Congress on endangered species, I'm always asked, "Can't we safely reduce the spotted owl to small numbers, keeping some in captivity as insurance?" The meaning is clear: "Let's log out almost all of western North America's old-growth forests, because if we can save species with high-tech solutions the forest doesn't matter." Let's tolerate a high risk of extinction.

Conservation is about the ecosystems that species define and on which they depend. It's about finding alternative, sustainable futures for peoples, for forests, and for wetlands. Molecular gimmickry does not address these core problems.

We shouldn't limit science. I, too, celebrate its successes. The idea we should retire is that new technically clever solutions suffice to fix our world. Common sense is necessary.

SCIENTISTS SHOULD STICK TO SCIENCE

BUDDHINI SAMARASINGHE
Molecular biologist

It's a statistical fact that you're more likely to die while horseback riding (one serious adverse event every 350 or so exposures) than from taking Ecstasy (one serious adverse event every 10,000 or so exposures). Yet in 2009 the scientist who said this was fired from his position as chairman of the U.K.'s Advisory Council on the Misuse of Drugs. Professor David Nutt's remit was to make scientific recommendations to government ministers on the classification of illegal drugs based on the harm they can cause. He was dismissed because his statement highlighted how the U.K. government's policies on narcotics are at odds with scientific evidence. Today, the medical use of drugs such as cannabis remains technically illegal.

Such incidents of silencing are sadly commonplace when it comes to politically controversial scientific topics. The U.S. government muzzled climate scientists in a similar manner in 2007, when it was reported that 46 percent of 1,600 surveyed scientists were warned against using terms like "global warming" and 43 percent said their published work had been revised in ways that altered their conclusions. U.S. preparations for oncoming climate change were blocked as a result, a failing that persists today.

Going back further, the story of Nikolai Vavilov is chilling. Vavilov was a plant geneticist in the Soviet Union under Joseph Stalin. He was jailed in 1940 for criticizing the pseudoscientific views of Trofim Lysenko, a protégé of Stalin. Vavilov died of starvation in prison a few years later; scientific dissent from Lysenko's theories of Lamarkian inheritance was outlawed in 1948. Soviet agriculture languished for decades because of Lysenkoism; meanwhile, famine decimated the population.

The scientific method is defined by the Oxford English Dictionary as "a method or procedure . . . consisting in systematic observation, measurement, and experiment, and the formulation, testing, and modification of hypotheses." It's our finest instrument for unearthing the truth. Applied correctly, it's blind to and corrects for our inherent biases. Scientists are trained to wield this formidable tool in their quest to understand the universe around us. The truths they uncover can be at odds with our current beliefs, but when the facts (based on evidence and arrived at through rigorous testing) change, minds also need to change.

I use the examples above of sidelined scientists to illustrate the consequences of excluding science from the policy-making process. But sometimes the sidelining is self-imposed; scientists can be genuinely reluctant to get involved in such activity and instead prefer to focus on gathering data and publishing results.

There's a tacit understanding, a custom in the culture of science, that scientists practice the scientific method in the confines of the ivory tower. Scientists are seen as impartial, aloof individuals with a single-minded focus on their work and out of touch with the realities of the world around them. They're expected to only do science, to find the truth and then leave it up to everyone else to decide what to do with it.

This is untenable. Scientists have a moral obligation to engage with the public about their findings—to advise and speak out on policy and critique its implementation. Science affects the life of every single species on our planet. It's ludicrous that the very people who discover the facts are not part of any subsequent policy-making dialog. Science needs to be an essential component of the public discourse; currently it is not. The consequences of that disconnect can be dire, as evinced by the criminalization of drugs that can provide relief to sufferers of chronic pain, troubling delays in programs of vital national importance, and the famine that slaughtered millions of Soviet citizens under Stalin's regime.

Scientists shouldn't simply stick to doing science. Perhaps we need to extend the scientific method to include a requirement for communication. Young scientists should be taught the value and necessity of communicating their findings to the general public. Scientists shouldn't shy away from controversy, because some topics shouldn't be controversial to begin with. The scientific evidence for the efficacy of vaccines, the process of evolution, the existence of anthropogenic climate change—all are accepted in the scientific community. Yet within the public sphere, goaded by a sensationalizing mainstream media and politicians seeking reelection, these settled facts are made to appear tentative. Science is based on evidence, and if that evidence tells us something new, we need to incorporate that into our policies. We cannot ignore it simply because it's unpopular or inconvenient.

By passionately advocating for evidence-based policy, scientists will expand scientific research, reversing the trend of recent years, and by thus visibly working for the common weal, they will earn the public's trust, protecting long-term investigations from short-sighted cuts. Scientific advancement is dependent on public funding and public backing. The space race, the Human Genome Project, the search for the Higgs boson, and the Mars *Curiosity* rover mission were all enthusiastically embraced by the public. The progress of science demands that scientists engage the public. But for that to happen, the notion that a scientist should stay hidden away in a laboratory needs to be retired.

NATURE = OBJECTS

SCOTT SAMPSON

Vice president of research and collections, Denver Museum of Nature and Science; dinosaur paleontologist; author, Dinosaur Odyssey

One of the most prevalent ideas in science is that nature consists of objects. Of course, the very practice of science is grounded in objectivity. We objectify nature so that we can measure it, test it, and study it, with the ultimate goal of unraveling its secrets. Doing so typically requires reducing natural phenomena to their component parts. Most zoologists, for example, think of animals in terms of genes, physiologies, species, and the like.

Yet this pervasive, centuries-old trend toward reductionism and objectification tends to prevent us from seeing nature as subjects, though there's no science to support such myopia. On the contrary, to give just one example, perhaps the deepest lesson cascading from Darwin's contributions is that all life on Earth, including us, arose from a single family tree. To date, however, this intellectual insight has yet to penetrate our hearts. Even those of us who fully embrace the notion of organic evolution tend to regard nature as resources to be exploited rather than relatives deserving of our respect.

What if science were to conceive of nature as both object and subject? Would we need to abandon our cherished objectivity? Of course not. Despite their chosen field of study, the vast bulk of social scientists don't struggle to form emotional bonds with family and friends. More so than at any point in the history of science, it's time to extend this subject/object duality to at least the nonhuman life-forms with which we share this world.

Why? Because much of our unsustainable behavior can be traced to a broken relationship with nature, a perspective that treats the nonhuman world as a realm of mindless, unfeeling objects. Sustainability will almost certainly depend on developing mutually enhancing relations between humans and nonhuman nature. Yet why would we foster such sustainable relations unless we cared about the natural world?

An alternative worldview is called for, one that reanimates the world. This mindshift, in turn, requires no less than the subjectification of nature. The notion of nature-as-subjects is not new; indigenous peoples around the globe tend to view themselves as embedded in animate landscapes replete with relatives. We have much to learn from this ancient wisdom.

To subjectify is to interiorize, such that the exterior world interpenetrates our interior world. Whereas the relationships we share with subjects often tap into our hearts, objects are dead to our emotions. Finding ourselves in relationships, the boundaries of self can become permeable and blurred. Many of us have experienced such transcendent feelings during interactions with nonhuman nature, from pets to forests.

But how might we undertake such a grand subjectification of nature? After all, worldviews become deeply ingrained; they're like the air we breathe—essential but ignored.

Part of the answer is likely to be found in the practice of science itself. The reductionist Western scientific tradition has concentrated on the nature of substance, asking, "What is it made of?" Yet a parallel approach—also operating for centuries, though often in the background—has investigated the science of pattern and form. Generally tied to Leonardo da Vinci, the latter method has sought to explore relationships, which can be notoriously difficult to quantify and must instead be mapped. The science of patterns has seen a recent resurgence, with abundant attention directed toward such fields as ecology and complex adaptive systems. Yet we've only scratched the surface, and much more integrative work remains to be done that could help us understand relationships.

Another part of the answer is to be found in education. We need to raise our children so that they see the world with new eyes. At the risk of heresy, I'd argue that science education, in particular, could be reinvigorated with subjectification in mind. Certainly the practice of science—the actual *doing* of scientific research—must be as objective as possible. But the *communication* of science could be done using both objective and subjective lenses.

Imagine if the bulk of science education took place outdoors, in direct, multisensory contact with the natural world. Imagine if students were encouraged to develop a meaningful sense of place through an understanding of the deep history and ecological workings of that place. And imagine if mentors and educators emphasized not only the identification and functioning of parts (say, of flowers or insects) but also the notion of organisms as sensate beings in intimate relationships with one another (and us). What if students were asked to spend more time learning about how a particular plant or animal experiences its world?

In this way, science—and biology, in particular—could help bridge the chasm between humans and nature. Ultimately, science education, in concert with other areas of learning, could go a long way toward achieving the "Great Work" described by cultural historian Thomas Berry—transforming the perceived world "from a collection of objects to a communion of subjects."

SCIENTIFIC MORALITY

EDWARD SLINGERLAND

Professor of Asian studies, Canada Research Chair in Chinese Thought and Embodied Cognition, University of British Columbia; author, Trying Not to Try

Impressed by the growing explanatory power of the natural sciences of his time, the philosopher David Hume called on his colleagues to abandon the armchair, turn their attention to empirical evidence, and "hearken to no arguments but those which are derived from experience . . . [to] reject every system of ethics, however subtle or ingenious, which is not founded on fact and observation." This was more than 250 years ago, and unfortunately not much changed in academic philosophy until about the last decade or two. Pushing past a barrier also associated with Hume—the infamous is-ought or fact-value distinction—a growing number of philosophers have finally begun arguing that our theories should be informed by our best current empirical accounts of how the human mind works, and that an ethical system that posits or requires an impossible psychology should be treated with suspicion.

One of the more robust and relevant bits of knowledge about human psychology that has emerged from the cognitive sciences is that we're not rational minds housed in irrational, emotional bodies. Metaphors like that of Plato's rational charioteer bravely struggling to control his irrational, passionate horses appeal to us because they map well onto our intuitive psychology, but they turn out to be ultimately misleading. A more empirically accurate image would be that of a centaur:

Rider and horse are one. To the best of our knowledge, there's no ghost in the machine. We're thoroughly embodied creatures, embedded in a complex social and culturally shaped environment, primarily guided in our daily lives not by cold calculation but hot emotion; not conscious choice but automatic, spontaneous processes; not rational concepts descended from the realm of Forms but modal, analogical images.

So, the ironic result of adopting a scientific stance toward human morality is to lay bare the impossibility of a purely scientific morality. The thoroughly rational, evidence-guided utilitarian is as much of a myth as the elusive *Homo economicus*, and equally as worthy of our disdain. Evolution may be utilitarian, guided solely by considerations of costs and benefits, but the ruthlessly utilitarian process of biocultural evolution has produced organisms that are, at a proximate level, incapable of functioning in a completely utilitarian fashion, and for very good design reasons. Because of rational evolutionary considerations, we cannot help but react irrationally to unfair offers in the Ultimatum Game, challenges to our honor, or perceived threat to our loved ones or cherished ideals. We're culturally infused animals guided largely by automatic habits, barely conscious hunches, profoundly motivating emotions, and wholehearted commitment to spooky, nonempirical entities ranging from human rights to the Word of God to the coming proletarian Utopia.

Science is powerful and important because it represents a set of institutional practices and thinking tools that allow us, *qua* scientists or intellectuals, to bootstrap ourselves out of our immediate perceptions and proximate psychology. We can understand that the Earth goes around the sun, that wonderful design can be the product of a blind watchmaker, or that the human mind is, in an important sense, reducible to biological processes. This gives us some helpful leverage over our evolved psychology, and I join many in thinking that this more accurate knowledge about ourselves and our world might allow us to devise—and maybe embrace—novel ethical commitments that could lead to more satisfying lives and a more just world. But let's not lose sight of the fact that science cannot bootstrap us out of our evolved minds themselves. The desire to bring about a more equitable, fair, and peaceful world is itself an emotion, an ultimately irrational drive grounded in commitment to ideals like human dignity, freedom, and well-being which we have inherited—in stripped-down, theologically minimalist form—from the cultural/religious traditions into which we were born. In their latest, liberal iterations, these ideals are rather odd—very few cultures have embraced diversity and tolerance as ethical desiderata, for instance—and are far from being universally embraced even in our contemporary world.

So, the myth that we secular liberals have emerged into a neutral place where we stand freed of all belief and superstition, guided solely by rationality, evidence, and clearly perceived self-interest, is something that needs to be retired. It's simply not the case that secular liberalism, grounded in materialist utilitarianism, is the inevitable and default worldview of anyone who isn't stupid, brainwashed, or uneducated; this idea seriously impedes our ability to understand people in earlier historical periods, from other cultures, and even ourselves.

Acknowledging that doesn't entail wallowing in postmodern relativism or blindly marching to a fundamentalist beat. Scientific inquiry, in its broad sense, is so effective at giving us reliable information about the world that to seriously defend any other method of inquiry as superior—or even equal—is perverse. There's also arguably a pragmatic case to be made that secular liberalism is the best worldview humans have ever come up with, or at least that individuals, given a choice, tend to preferentially gravitate toward it. In any case, it's our value system, and the very nature of evolved human psychology makes it impossible for us not to want to defend human dignity or women's rights and, when appropriate, impose them on others. But recognizing the limitations of reason allows us to articulate and defend such values in a more effective way. It also allows us to better understand, scientifically, problems such as the causes of religious violence or the roots of persistent international conflicts, or moral challenges such as balancing our folk intuitions about personal responsibility with a neuroscientific understanding of free will. The science of morality requires us to, in the end, get beyond the myth of a perfectly objective scientific morality.

SCIENCE IS SELF-CORRECTING

ALEX HOLCOMBE

Associate professor of psychology, University of Sydney; codirector, Centre for Time; associate editor, Perspectives on Psychological Science

The pace of scientific production has quickened, and self-correction has suffered. Findings that might correct old results are considered less interesting than results from more original research questions. Potential corrections are also more contested. As the competition for space in prestigious journals has become increasingly frenzied, doing and publishing studies that would confirm the rapidly accumulating new discoveries, or would correct them, has become a losing proposition.

Publication bias is the tendency to not publish "negative," or nonconfirmatory, results. Its effect, the suppression of corrections, can prevail even when much work has gone into obtaining the negative results. Ideally, such findings would move quickly and easily from individual scientists' laboratories to public availability. But the path can be so difficult, and is so infrequently used, that many areas of science don't deserve the self-correcting moniker.

The prestigious journals in many fields make no bones about it, declaring that they're in the business of publishing exciting discoveries that advance the field in new ways, not studies similar to previous ones which find a less interesting result. Even at those journals claiming to welcome negative findings, a would-be corrector faces an uphill battle. The scientists who vet the new evidence for the journal typically include the author(s) of the original and possibly incorrect conclusion.

Human frailty, egotism, and anonymity together bias reviewers' verdicts toward "reject." That's normally enough to deny new negative results an appearance in a journal.

Self-correction is thus undermined by several factors, some human and some institutional. The institutional factors are sometimes historical accidents. One is the number of venues where it's considered appropriate to publish your work. In certain subfields, almost all new work appears in only a very few journals, all associated with a single professional society. There's no way around the senior gatekeepers, who can suppress corrections with impunity. Fields with a variety of publication venues and stakeholders are healthier; it's more difficult for a single school of thought to take over.

Several fields, such as astrophysics, have a culture of sharing and citing manuscripts before they're even submitted to journals. Researchers need only post their manuscript to a website, such as arXiv.org. The result reported might then be ignored but cannot be fully suppressed. In principle, all areas of science could adopt this practice, but for now most stick with their secretive reviews, which frequently torpedo new results.

The bias against corrections is especially harmful in areas where the results are cheap but the underlying measurements are noisy. In those scientific realms, the literature may quickly become polluted with statistical flukes. Unfortunately, these two features—cheap results and noisy measurements—are characteristic of most sub-areas of psychology, my own discipline. Some other fields, such as contemporary epidemiology, may have it even worse—particularly with regard to a third exacerbating factor, the small size of the true effects investigated. As Stanford professor of medicine John Ioannidis has pointed out, the smaller the true effects in an area, the more likely it is that a given claimed effect is instead a statistical fluke (a false positive).

There are fixes, one of which both improves the behavior of individual researchers and dissolves institutional obstacles: public registration of the design and analysis plan of a study before it's begun. Clinical-trials researchers have done this for decades, and in 2013 researchers in other areas followed suit. Registration includes the details of the data analyses that will be conducted, which eliminates the former practice of presenting the inevitable fluctuations of multifaceted data as robust results. Reviewers assessing the associated manuscripts end up focusing on the soundness of the study's registered design rather than disproportionately favoring the findings. This helps reduce the disadvantage that confirmatory studies usually have relative to fishing expeditions. Indeed, a few journals have begun accepting articles from well-designed studies even before the results come in.

The Internet's explosive growth has led to pervasive public ratings of, and useful comments on, nearly every product and service—but somehow not for scientific papers, in spite of the obvious value of commenting for pointing out flaws and correcting errors. Until recently, to point out a problem with a paper in a place other researchers would come across, you had to run the gauntlet of the same editors and reviewers who had missed or willfully overlooked the problem in the first place. Those reviewers, as experts also publishing in the area, frequently have commitments, just as the authors do, to flawed practices or claims. Now, finally, scientists are taking advantage of the Internet to contribute expertise and opinions that go beyond the authors of an article and its two or three reviewers. In October 2013, the U.S. National Library of Medicine began allowing researchers to post comments on practically any paper in biology and medicine, via PubMed, the most widely used database of such papers. Correction of simple errors is no longer arduous.

Besides simple error correction, comments can provide new perspectives. Cross-pollination of ideas then increases. Exhausted research areas will be revitalized by the introduction of new approaches, and attacks by researchers from outside a field will break hardened orthodoxies. But hiring, promotion, and grant committees typically don't value the contributions made by individual researchers using these tools. As long as this continues, progress may be slow. As Max Planck observed, revolutions in science sometimes have to wait for funerals. Even after the defenders of old practices go to their final resting places, the antiquated traditions sometimes endure, in part from the support of institutional policies. A policy doesn't die until someone kills it. Reforms and innovations need our active support; only then can science live up to its self-correcting tagline.

REPLICATION AS A SAFETY NET

ADAM ALTER

Psychologist; assistant professor of marketing, Stern School of Business, NYU; author, Drunk Tank Pink: And Other Unexpected Forces That Shape How We Think, Feel, and Behave.

In 1984, New York became the first state to introduce mandatory seat-belt laws. Most of the remaining states applauded the new legislation and followed suit in the 1980s and 1990s, but a small collection of researchers worried that seat belts might paradoxically license people to drive less carefully. They believed that people drove carefully because they worried about being seriously injured in an accident; if seat belts diminished the risk of serious injury, they would also diminish the incentive to drive carefully.

There's a danger that social scientists rely too heavily on the concept of replication just as potentially careless drivers rely too heavily on seat belts. When we examine new hypotheses, we tolerate the possibility that approximately one in every twenty results is a fluke. If we run the experiment two or three times, and the result is replicated, it's safer to assume the original result was reliable. Students are taught that untruths will be revealed in time through replication—that flimsy results will wither under empirical scrutiny, so the enduring scientific record will reflect only those results that are robust and replicable. Unfortunately, this appealing theory crumbles in practice.

As the seat-belt illustration suggests, the problem begins when researchers behave carelessly because of overreliance on the theory of replication. Each experiment becomes less valuable and less definitive, so instead of striving to craft the cleanest, most informative experiment, the incentives weigh in favor of running many unpolished experiments instead.

Journals are, similarly, more inclined to publish marginally questionable research because they expect other researchers to test the genuineness of the effect. But in fact only a limited sample of high-profile findings are replicated, because there's less scientific glory in overturning an old finding than in producing a new one. Given limited time and resources, researchers tend to focus on testing new ideas rather than on questioning old ones. The scientific record features thousands of preliminary findings but relatively few thorough replications, rejoinders, and reconsiderations of those early results.

Without a graveyard of failed effects, it's difficult to distinguish robust results from brittle flukes. The serious consequence is that our overreliance on the theory of replication—the notion that researchers will unmask empirical untruths—means that we overestimate the reliability of the many effects that have yet to be reexamined. Replication is a critical component of the scientific process, but the illusion of replications as an antidote to flimsy effects deserves to be shattered.

SCIENTIFIC KNOWLEDGE STRUCTURED AS "LITERATURE"

BRIAN CHRISTIAN

Author, The Most Human Human: What Artificial Intelligence Teaches Us About Being Alive

What's most outmoded within science, most badly in need of retirement, is the way we structure and organize scientific knowledge itself. Academic literature, even as it moves online, is a relic of the era of typesetting, modeled on static, irrevocable, toothpaste-out-of-the-tube publication. Just as the software industry has moved from a "waterfall" process to an "agile" process—from monolithic releases shipped from warehouses of mass-produced disks to over-the-air differential updates—so must academic publishing move from its current read-only model and embrace a process as dynamic, up-to-date, and collaborative as science itself.

It amazes me how poorly the academic and scientific literature is configured to handle even retraction, even at its most clear-cut—to say nothing of subtler species like revision. Typically (for example), even when the journal editors *and* the authors fully retract a paper, the paper continues to be available at the journal's website and with no indication that a retraction exists elsewhere, let alone on the same site, penned by the same authors and vetted by the same editor. (Imagine if the Food and Drug Administration allowed a drug maker to continue manufacturing a drug known to be harmful as long as they also manufactured a warning label—but were under no obligation to put the label on the drug.)

A subtler question is how and in what manner (*caveat lector?*) to flag studies that depend on the discredited study (let alone studies depending on those studies). Citation is the obvious first answer, although it's not quite enough. In academic journals, all citations attest to the significance of the works they cite, regardless of whether their results are being presumed, strengthened, or challenged; even theories used as punching bags are accorded the respect of being worthy or significant punching bags. But academic literature makes no distinction between citations merely considered significant and ones additionally considered true. What academic literature needs goes deeper than the view of citations as kudos and shout-outs. It needs what software engineers have used for decades: *dependency management.*

A dependency graph would tell us, at a click, which of the pillars of scientific theory are truly load-bearing. And it would tell us, at a click, which other ideas are likely to get swept away with the rubble of a particular theory. Academic publishers worth their salt could, for instance, flag not only retracted articles (that this isn't standard practice is, again, inexcusable) but articles that depended in some meaningful way on the results of the retracted work.

Academic publishers worth their salt would also accommodate another pillar of modern software development: *revision control.* Code repositories, like wikis, are living documents, open not only for scrutiny, censure, and approbation but also for modification. In a revision-control system like Git (and its wildly successful open-source community on GitHub), users can create "issues" that flag problems and require the author's response. They can create "pull requests" that propose answers and alterations, and they can "fork" a repository if they want to steward their own version of the project and take it in a different direction. (Sometimes forked repositories serve a niche audience; sometimes they wither from neglect or disuse; sometimes they fully steal the audience and user base from the original; sometimes the two continue to exist in parallel or continue to diverge; and sometimes they're reconciled and reunited downstream.) A Git repository is the best of top-down and bottom-up, of dictatorship and democracy; its leaders set the purpose and vision and have ultimate control and final say—yet any citizen has an equal right to complain, propose reform, start a revolt, or simply pack his bags and found a new nation next door.

The "Accept," "Reject," and "Revise and Resubmit" ternary is anachronistic, a relic of the era of metal type. Even peer review itself, with its anonymity and bureaucracy, may be ripe for reimagining. The behind-closed-doors, anonymous review process might be replaced with, say, something closer to a "beta" period. The article need not be held up for months—at least, not from other researchers—while it's considered by a select few. One's critics need not be able to clandestinely delay one's work by months. Authors need not thank "anonymous readers who spotted errors and provided critical feedback" when those readers' corrections are directly incorporated (with attribution) as differential edits. Those readers need not offer their suggestions as an act of obligation or charity, and they need not go unknown.

Some current rumblings of revolution seem promising. Wide circulation among academics of "working papers" challenges the embargo and lag in the peer-review process. PLOS ONE insists on top-down quality assurance but lets importance emerge from the bottom up. Cornell's arXiv project offers a promising alternative to more traditional journal models, including versioning (and its "endorsement" system has, since 2004, suggested a possible alternative to traditional peer reviews). However, its interface, by design, limits its participatory and collaborative potential.

On that front, a massive international collaboration via the Polymath Project Web site in 2013 successfully extended the work of Yitan Zhang on the twin primes conjecture (and I understand the University of Montreal's James Maynard has subsequently gone even further). Amazingly, this groundbreaking collaborative work was done primarily in a comment thread.

The field cries out for better tools; meanwhile, better tools already exist in the adjacent field of software development. It's time for science to go agile. The scientific literature, taken as content, is stronger than it has ever been—as, of course, it should be. As a form, the scientific literature has never been more inadequate or inept. What's in most dire need of revision is revision itself.

THE WAY WE PRODUCE AND ADVANCE SCIENCE

KATHRYN CLANCY
Assistant professor of anthropology,
University of Illinois, Urbana-Champaign

Last year, I spearheaded a survey and interview research project on the experiences of scientists at field sites. Over 60 percent of the respondents had been sexually harassed, and 20 percent had been sexually assaulted. Sexual predation was only the beginning of what I and my colleagues uncovered: Study respondents reported psychological and physical abuses, like being forced to work late into the day without being told when they could head back to camp, not being allowed to urinate, verbal threats and bullying, and being denied food. The majority of perpetrators are fellow scientists senior to the targets of abuse, the targets usually being female graduate students. Since we started analyzing these data, I haven't been able to read a single empirical science paper without wondering on whose backs, via whose exploitation, that research was conducted.

When the payoff is millions of dollars of research money, *New York Times* coverage, Nobel Prizes, or even just tenure, we often seem willing to pay any price for scientific discovery and innovation. This is exactly the idea that needs to be retired—that science should be privileged over scientists.

Putting ideas above people is a particularly idealistic way of viewing the scientific enterprise. This view assumes that the field of science is not only meritocratic but that who a scientist is, or where she comes from, plays no role in her level of success. Yet it's well known that class and occupational and educational attainment vary by race, gender, and many other aspects of human diversity, and that these factors do influence who chooses a scientific career and who stays in science. As unadulterated as we may want to envision science, the scientific enterprise is run by people, and people often run on implicit bias. Scientists know these things—scientists wrote the papers to which I refer—but I'm not sure we have all internalized the implications. The implications for implicit bias and workplace diversity are that social structure and identity motivate interactions between workers, increasing the chances for exploitation in terms of both overwork and harassment—particularly for those who are junior or underrepresented.

Scientists aren't blind to the problems of the ways we culturally conceive of scientific work. There are increasing discussions among them of the ever elusive work/life balance. By and large, these conversations center around personal ways we can create a better life for ourselves through management of our time and priorities. To my mind, these conversations are a luxury for those who have already survived the gauntlet of being a trainee scientist. There are few ways to consider or improve work/life balance when you're one of the grunts on the lab floor or fossil dig.

Overwork and exploitation don't lead to scientific advancement nearly as effectively as humane, equitable, and respectful workplaces. For instance, recent social-relations modeling research reveals that when women are integrated rather than peripheral members of their laboratory group, those labs publish more papers. Further, years of research on counterproductive work behaviors demonstrates that when you create strongly enforced policies and independent lines of reporting, work environments improve and workers are more productive. The hassled, overworked, give-it-all-for-the-job mentality in science has not been found empirically to produce the best work.

The lives of scientists need to be prioritized over scientific discovery in the interest of doing better science. Many of us operate on fear—fear of being scooped, fear of not getting tenure, fear of not having enough funding to do our work, fear even of being exploited ourselves. But we cannot let fear motivate a scheme that crushes potential bright future scientists. The criteria for scholarly excellence shouldn't be based on who survives or evades poor treatment but on who has the intellectual chops to make the most meaningful contributions. Thus, trainees need unions and institutional policies to protect them, and senior scientists need to enact cultural change. An inclusive, humane workplace is the one that will lead to the most rigorous, world-changing scientific discoveries.

ALLOCATING FUNDS VIA PEER REVIEW

AUBREY DE GREY

Gerontologist; chief science officer, SENS Foundation; author, Ending Aging

From top to bottom of the profession, scientists are forsaking their chosen vocation in greater numbers than ever before, in favor of a more dependable and less stressful source of income. What is the basis of this stress and uncertainty which so severely depletes the ranks of that indispensable community who seek to further humanity's understanding of nature and thereby our ability to manipulate nature for the greater good? At the sharp end, it's the members of those ranks—scientists themselves—via the convention of apportioning funding by peer review of grant applications.

Only at the sharp end, of course. I certainly don't lay blame at scientists' feet. In fact, I don't really lay blame anywhere: The issue is that the prevailing system evolved in a different time, and in circumstances to which it was well suited, but has signally failed to adapt—indeed, has shown itself intrinsically nonadaptable—to present conditions. What's needed is a replacement system that solves the problems everyone in science agrees exist today but still distributes funds according to metrics that all constituencies agree is fair.

The peer-review system is apparently a local maximum: Numerous tweaks have been proposed, but all have resisted adoption because they do more harm than good. But is it a *global* maximum? Is it, as Churchill described democracy, the worst option except for all the others? Or could a radical departure rank higher, by all key measures? Here I sketch a possible option. I'm not sure it ticks all the boxes, but it shows sufficient promise as a candidate that the scientific community need no longer bear with the current system on the assumption that nothing better is possible.

First, briefly: What's so wrong with peer review of grant applications these days? Two words: "pay line." Peer review evolved when the balance between supply of and demand for public research funds was such that at least 30 percent of applications could be funded. It worked well: If you didn't really know how to design a project or how to communicate its value to your colleagues or how to perform it economically, these failings would emerge and you'd learn how to avoid them, until eventually those colleagues would recommend to the government that you be given your chance. But these days the corresponding percentage is typically in single digits. Does that mean you just have to be really good? I wish.

What it actually means is that you have to be not only really good but also really persistent, and moreover—and this is by far the worst aspect—really, really convincing in your argument that the project will succeed. What's so bad about that? Simply that some projects are (much) easier than others, and the hard ones tend to be those that determine the long-term rate of progress of a discipline, even though they have a significant failure rate. A system that neglects high-risk/high-gain work hugely slows scientific progress, with catastrophic consequences for humanity. Cross-disciplinary research—work drawing together ideas not previously combined, which historically has also been exceptionally fruitful—is almost impossible to get funded, simply because no research panel ("study section," in NIH vernacular) has the necessary range of expertise to understand the proposal's full value.

All this would be largely solved by a system based on peer recognition rather than peer review. When a scientist first applies for public research funds, his or her career will be divided into five-year periods, starting with the past five years (period 0), the coming five (period 1), and so on. Period 1 is funded at a low, entry-level rate on the basis of simple qualifications (possession of a doctorate, number of years of post-doctoral study, etc.), and *without the researcher having provided any description of what specific research is to be undertaken.* Period 2's funding level is determined, as a percentage of total funds available for the scientist's chosen discipline, again without any description of work to be performed but instead on the basis of how well cited was the work performed in period 0.

This decision is made at the end of period 1 year 4, based on all citations since period 0 year 2 (so a total of eight years) to papers published in period 0 year 2 through period 1 year 1 (five years, approximating the interval when work done during period 0 will have been published). Citations are weighted according to whether the applicant is a first/senior/middle author; self-citations aren't counted; only papers reporting new research that depended on research funds are counted. Consideration is given to seniority and level of funding during the relevant period, according to a formula applied across the board rather than by discretion. Funding for period 3 is determined similarly, at the end of period 2 year 4, on the basis of work performed during period 1, and so on. Flexibility is incorporated concerning front-loading of funds to year 1 of a given period, to allow for large capital expenditures.

This scheme improves on the current system in many ways.

Zero time is spent preparing and submitting (and resubmitting . . .) descriptions of proposed research, and zero money is spent on evaluating such proposals. Bias against high-risk/high-gain work is greatly reduced, both by the absense of peer review and also because funding periods exceed the currently typical three years. Significance of past work is evaluated after an appropriate period of time, not by such first-impression measures as the impact factor of journals where one has just published. One can also split one's application across multiple disciplines, with funding level from each discipline proportioned accordingly, removing the bias against cross-disciplinary research. Finally, one has a year at the end of a period to plan what work will be done in the next period, in full knowledge of what resources will be at one's disposal.

Researchers are of course free to seek additional funds from elsewhere—and indeed, some public funds could still be apportioned via the traditional method. Thus this option need not represent a particularly radical departure. It could easily be phased in. Worth considering?

SOME QUESTIONS ARE TOO HARD FOR YOUNG SCIENTISTS TO TACKLE

ROSS ANDERSON

Professor of security engineering, Cambridge University; author, Security Engineering

Max Planck famously described the progress of science as being "one funeral at a time," as the old-school physicists died off and their jobs were taken by young men who followed the new quantum religion.

This brutal style of scientific revolution has left some rather rigid scar tissue. For many years, it has been almost taboo to suggest that the questions at the foundations of quantum mechanics might actually have an answer. Yet new results in different areas of physics, chemistry, and engineering are beginning to suggest that there might possibly be an answer after all.

At the Solvay Conference in 1927, Niels Bohr and Werner Heisenberg out-debated Albert Einstein and Louis de Broglie; they persuaded the world that we should just take the tools of the new quantum mechanics on faith rather than trying to derive them from underlying classical principles. This Copenhagen school, the "shut up and calculate" school of quantum mechanics, rapidly became the orthodoxy. It was reinforced when calculations by John Bell were experimentally verified by Alain Aspect in 1982 and appeared to show that reality at the quantum level could not be both local and causal.

While some philosophers of physics toyed with exotic interpretations of quantum mechanics, most physicists shrugged;

they accepted quantum foundations as a "certified insoluble" problem and told their graduate students not to even think about wasting their lives on looking for a solution. Others just loved the idea that physics proves the world is too complex to understand and that the proof is beyond the comprehension of outsiders. Physicists could be the new high priests as the quantum became the core magic. Recently we've got quantum with everything, from cryptography to biology; the word has become a magic spell for fund-raising. So long as no one dared challenge this for fear of being thought a crank or dismissed as an outsider, we were stuck.

Things are starting to change. In physics, Yves Couder and Emmanuel Fort found that bouncing droplets on a bath of vibrating oil mimic many phenomena previously thought unique to the quantum world, including single-slit and double-slit refraction, tunneling, and quantized energy levels. In chemistry, Masanao Ozawa and Werner Hofer have shown that the uncertainty principle is only approximately true: Modern scanning probe microscopes can often measure the position and momentum of atoms slightly more accurately than Heisenberg predicted—which should worry people who claim that quantum cryptography is "provably" secure! In computing, the promised quantum computers are still stuck at factoring 15, despite hundreds of millions in research funding over almost twenty years. And the physicist Theo van Nieuwenhuizen has pointed out a contextuality loophole in Bell's theorem that looks rather hard to fix.

There's a striking parallel with another big problem in science—consciousness. For years, the few first-division academics who dared tackle such problems tended to be near retirement and famous enough to shrug off disapproval. Just as

Daniel Dennett and Nicholas Humphrey wrote on consciousness, Tony Leggett and Gerard 't Hooft wrote on quantum foundations—so the flame was kept alight. But it's time to bring some tinder. Viennese physicists have now organized two symposia on emergent quantum mechanics, as people finally dare to wrestle with what might be going on down there.

So the idea I'd like to retire is the idea that some questions are just too big for normal working scientists to tackle. Old-timers should not try to erect taboos around the problems that have eluded us. We must cheerfully challenge the young: "Prove us wrong!" As for young scientists, they should dare to dream and to aim high.

ONLY SCIENTISTS CAN DO SCIENCE

KATE MILLS

Doctoral student, Institute of Cognitive Neuroscience, University College London

Currently, most individuals funded or employed to conduct scientific experiments have been trained in traditional academic settings. This includes not only the twelve years of compulsory education but also another six to ten years of university education, often followed by years of postdoctoral training. While this formal academic training undoubtedly equips one with the tools and resources to become a successful scientist, informally trained individuals of all ages are just as able to contribute to our knowledge of the world through science.

These "citizen scientists" are often lauded for lightening the load on academic researchers engaged in Big Data projects. Citizen scientists have contributed to these projects by identifying galaxies or tracing neural processes, and typically without traditional incentives or rewards, like payment or authorship. However, limiting the potential contributions of informally trained individuals to data collecting or data processing discounts the abilities of citizen scientists to inform study design, data analysis, and interpretation. Soliciting the opinions of individuals who are participants in scientific studies (e.g., children, patients) can help traditional scientists design ecologically valid and engaging studies. Equally, these populations might have their own scientific questions or provide new and diverse perspectives to the interpretation of results.

Importantly, science is not limited to adults. Children as young as eight have coauthored scientific reports. Teenagers have made important health discoveries with tangible outcomes. Unfortunately, these young scientists face many obstacles that institutionally funded individuals often take for granted, such as access to previously published scientific findings. While the rise of open-access publication, as well as many open-science initiatives, make the scientific environment friendlier for citizen scientists, many traditional science practices remain out of reach for those without sufficient funds.

What we think we know about ourselves through science could be skewed, since the majority of psychology studies sample individuals who don't represent the population as a whole. These WEIRD (Western, Educated, Industrialized, Rich, Democratic) samples make up most nonclinical neuroimaging studies as well. Increased awareness of this bias has prompted researchers to actively seek out samples that are more representative; however, there's less discussion or awareness of the potential biases introduced by WEIRD scientists.

If most funded and published scientific research is conducted by a sample of individuals trained to be successful in academia, then we're potentially biasing scientific questions and interpretations. Individuals who might not fit into an academic mold but nevertheless are curious to know the world through the scientific method face many barriers. Crowdfunded projects (and even scientists) are beginning to receive recognition from fellow scientists dependent on dwindling numbers of grants and academic positions. However, certain scientific experiments are more difficult, if not impossible, to conduct without institutional support—for example, studies involving human participants. Community-supported checks and balances remain essential for scientific projects, but perhaps they too can become unbound from traditional academic settings.

The means for collecting and analyzing data are becoming more accessible to the public each day. New ethical issues will need to be discussed and infrastructures built to accommodate those conducting research outside traditional settings. With this, we will see an increase in the number of scientific discoveries made by informally trained citizen scientists of all ages and backgrounds. Those previously unheard voices will add valuable contributions to our knowledge of the world.

THE SCIENTIFIC METHOD

MELANIE SWAN

Systems-level thinker; futurist; applied genomics expert; principal, MS Futures Group; founder, DIYgenomics

The scientific idea most ready for retirement is the scientific method itself. More precisely, it's the idea that there's only *one* scientific method, one exclusive way of obtaining scientific results. The traditional scientific method, as an exclusive approach, is not adequate to the new situations of contemporary science like Big Data, crowd-sourcing, and synthetic biology. Hypothesis-testing through observation, measurement, and experimentation made sense in the past, when obtaining information was scarce and costly, but this is no longer the case. In recent decades, we've been adapting to a new era of information abundance that has facilitated experimental design and iteration. One result is that there's now a field of computational science alongside nearly every discipline—for example, computational biology and digital manuscript archiving. Information abundance and computational advance have promulgated the evolution of a scientific model distinct from the traditional scientific method, and three emerging areas are advancing it even more.

Big Data, the creation and use of large and complex cloud-based data sets, is one pervasive trend reshaping the conduct of science. The scale is immense: Organizations routinely process millions of transactions per hour into 100-petabyte databases. Worldwide annual data creation is currently doubling and estimated to reach 8 zettabytes in 2015. Even before the Big Data era, modeling, simulating, and predicting became a key computational step in the scientific process, and the new methods required to work with Big Data make the traditional scientific method increasingly less relevant. Our relationship to information has changed with Big Data. Previously, in the era of information scarcity, all data was salient. In a calendar, for example, every data element, or appointment, is important and intended for action. With Big Data, the opposite is true; 99 percent of the data may be irrelevant (immediately, over time, or once processed into higher resolution). The focus becomes extracting points of relevance from an expansive whole, looking for signal from noise, anomalies, and exceptions—for example, genomic polymorphisms. The next level of Big Data processing is pattern recognition. High sampling frequencies allow not only point-testing of phenomena (as in the traditional scientific method) but its full elucidation over multiple time frames and conditions. For the first time, we can obtain longitudinal baseline norms, variance, patterns, and cyclical behavior. This requires thinking beyond the simple causality of the traditional scientific method into extended systemic models of correlation, association, and episode triggering. Some of the prominent methods used in Big Data discovery include machine-learning algorithms, neural networks, hierarchical representation, and information visualization.

Crowd-sourcing is another trend reshaping the conduct of science. This is the coordination of large numbers of individuals (the crowd) through the Internet to participate in some activity. Crowd models have led to the development of a science ecosystem that includes the professionally trained institutional researcher using the traditional scientific method at one end and the citizen-scientist exploring issues of personal interest through a variety of methods at the other. In between are different levels of professionally organized and peer-coordinated efforts. The Internet (and the trend to Internet-connect all people—2 billion, now estimated to be 5 billion in 2020) enables very-large-scale science. Not only are existing studies cheaper and quicker in crowd-sourced cohorts but studies 100 times the size and detail of previous studies are now possible. The crowd can provide volumes of data by automatically linking quantified self-tracking gadgets to data-commons websites. Citizen-scientists participate in light information-processing and other data collection and analysis activities through websites like Galaxy Zoo. The crowd is engaged more extensively through crowd-sourced labor marketplaces (initially like Mechanical Turk, now increasingly skill-targeted), data competitions, and serious gaming (like predicting protein-folding and RNA conformation). New methods for the conduct of science are being innovated through DIY efforts, the quantified self, biohacking, 3-D printing, and collaborative peer-based studies.

Synthetic biology is a third widespread trend reshaping the conduct of science. Lauded as the potential "transistor of the 21st century," given its transformative possibilities, synthetic biology is the design and construction of biological devices and systems. It's highly multidisciplinary, linking biology, engineering, functional design, and computation. One of the key application areas is metabolic engineering, working with cells to expand their usual production of substances, which can then be used for energy, agricultural, and pharmaceutical purposes. The nature of synthetic biology is proactively creating *de novo* biological systems, organisms, and capacities, which is the opposite of the passive characterization of phenomena for which the original scientific method was developed. While it's true that optimizing genetic and regulatory processes within cells can be partially construed under the scientific method, the overall scope of activity and methods are much broader. Innovating *de novo* organisms and functionality requires a significantly different scientific methodology than that supported by the traditional scientific method and includes a reconceptualization of science as an endeavor of characterizing *and* creating.

We can no longer rely exclusively on the traditional scientific method in the new era of science emerging in areas like Big Data, crowd-sourcing, and synthetic biology. Many models should be employed for the next generation of scientific advance, supplementing the traditional scientific method with new ways that are better suited and equally valid, and opening up new tiers for the conduct of science. Science can now be carried out downstream, at increasingly detailed levels of resolution and permutation, and upstream, with broader systemic dynamism. Temporality and the future become more knowable and predictable as all processes, human and otherwise, can be modeled with continuous realtime updates. Epistemologically, "how we know," and the truth of the world and reality, is changing. In some sense, we may be in an intermediate Dark Ages node, where the multiplicity of future science methods can pull us into a new Enlightenment just as surely as the traditional scientific method pulled us into modernity.

BIG EFFECTS HAVE BIG EXPLANATIONS

FIERY CUSHMAN

Assistant professor of psychology, Harvard University

Many scientists are seduced by a two-step path to success: First, identify a big effect and then find the explanation for it. There's an implicit theory behind this approach, which is that big effects have big explanations. And scientists are more interested in the explanations than in the effects: Newton is famous not for showing that apples and orbiting bodies both fall but for explaining why. So if the implicit theory is wrong, then a lot of people are barking up the wrong trees.

There is, of course, an alternative and plausible source of big effects: many small explanations interacting. As it happens, this alternative is worse than the wrong tree—it's a near hopeless tree. The wrong tree would simply yield a disappointingly small explanation. But the hopeless tree has so many explanations tangled in knotted branches that extraordinary effort is required to obtain any fruit at all.

So, do big effects tend to have big explanations or many explanations? There's probably no single, simple, and uniformly correct answer to this question. (It's a hopeless tree!) But we can use a simple model to make an educated guess.

Suppose the world is composed of three kinds of things. There are *levers* we can pull. Pulling these levers cause *observable effects* (lights flash, bells ring, apples fall). Finally, there's a hidden layer of *causal forces* (the explanations) that connect the levers to their effects.

In order to explore this toy world, I simulated it on my laptop. First, I created 1,000 levers. Each lever activated from 1 to 5 hidden mechanisms (200 levers activated just 1 mechanism each, another 200 activated 2, and so on). In my simulation, each mechanism was simply a number drawn from a normal distribution with a mean of zero. Then, the hidden mechanisms activated by each lever were summed to produce an observable effect. So 200 of the levers produced effects equal to a single number drawn from a normal distribution, another 200 levers produced effects equal to the sum of 2 such numbers, and so forth.

After this, I had a list of 1,000 effects of varying size. Some were large (very negative or very positive); others were small (close to zero). I looked at the 50 smallest effects, curious to see how many of them resulted from a single, isolated mechanism: 11 out of 50. Then I checked how many were the result of 5 mechanisms added together: 6 out of 50. The smallest effects tended to have fewer explanations.

Next, I looked at the 50 largest effects—about 100 times larger, on average—and found that they tended to have many more explanations. Twenty-five of them had 5 explanations; not one had a single explanation. The first such single-explanation effect was ranked 103rd in size. (These examples help make my point tangible, but its essence can be captured more succinctly: The standard deviation of the sum of two uncorrelated random variables is greater than the standard deviation of either individually).

So if a scientist's exclusive goal was simplicity, then in my toy world she ought to avoid the very biggest effects and instead pursue the smallest ones. Yet she might feel cheated, because this method would identify only explanations of little influence. As a crude method of balancing simplicity (few explanations) against influence (big explanations), I computed a sort of "expected value" of experimentation for different effect sizes: the probability of finding a one-cause effect multiplied by the size of the effect in question. As you might guess, the highest expected values tend to fall toward the middle of the range of effect sizes. Balance, it seems, finds a soul mate in modesty.

Now, there are some caveats to my back-of-the-envelope calculations. Most scientists are capable of working out causal mechanisms that have more than one dimension. (Some can even handle five!) Also, the actual causal mechanisms that scientists investigate are far more complicated than my model allows for. One explanation may be related to many effects; multiple explanations combine with each other nonlinearly; explanations may be correlated; and so forth.

Still, it's worth retiring the implicit theory that we should most doggedly pursue the largest effects. I suspect that every scientist has a favorite example of the perils of this theory. In my field, lakes of ink have been why people consider it acceptable to redirect a speeding trolley away from five people and toward one, but unacceptable to hurl one person in front of a trolley in order to stop it from hitting five. This case is alluring because the effect is huge and its explanation is not at all obvious. After years of research across many labs, however, there's considerable agreement that the effect doesn't have just one explanation. In fact, we've tended to learn more from studying much smaller effects with a key benefit, a sole cause.

It's natural to praise research that delivers large effects and the theories that purport to explain them. And this praise is often justified, not least because the world has large problems demanding ambitious scientific solutions. Yet science can advance only at the rate of its best explanations. Often the most elegant arise from effects of modest proportions.

SCIENCE = BIG SCIENCE

SAMUEL ARBESMAN

Applied mathematician; senior scholar, research and policy, Ewing Marion Kauffman Foundation; author, The Half-Life of Facts

Centuries ago, when science was young, it was possible to make contributions to scientific knowledge through simple experiments. You could be a hobbyist or a "gentleman scientist" and discover something fundamental about the world around us. But in the past several decades science has gotten bigger. In this era of Big Science, we need large teams of scientists working together to make discoveries, in everything from the life sciences to high energy physics. And we need lots of money to do this. The era of the lone scientist doing small-scale science seems to be over.

And that's often the narrative we hear. The Higgs boson wasn't discovered using an apparatus developed in a garage. It was found using a massive technological construction and thousands of scientists were involved.

So is small-scale science over? Although the trend is clearly toward team science, small and clever science—the realm of the tiny budget, the elegant experiment, sometimes even the hobbyist—is by no means defunct. Small science is not necessarily the lone underdog working against the establishment; more often it's simply one or two underfunded scientists doing their best. But it seems they survive even in this modern era of Big Science. Several years ago, for example, a paleontology graduate student cleared a dinosaur of cannibalism charges, a discovery that began with a very simple observation—by looking at one of the fossil casts on the wall of the American Museum of Natural History's subway station. Or consider the scientists who examined the space of possible ways to tie a necktie and whose research was published in *Nature*. Little science is still possible.

Though these examples might seem somewhat trivial, small-scale science can have a big impact. Peter Mitchell was awarded the 1978 Nobel Prize in chemistry for work he conducted at his own small private research institute with only a handful of people. Support for this lab included funds from his family—making Mitchell a modern-day equivalent of the gentleman scientist. Another Nobel Prize was awarded for work on split-brain patients—those with the connection between their two hemispheres severed—which led to novel insights into the brain's function. Part of this work consisted of experiments so simple (though exceedingly clever) that the Nobel website features a game just like the original experiments, which you can play at home.

You can still do science on the cheap. Several decades ago, Stanley Milgram measured the well-known six degrees of separation using little more than postcards. While science has become bigger since then, in some ways it has become even easier to conduct large-scale science by operating at a small scale. Thanks to huge computational advances and widespread data freely available (not to mention easier data collection online), any scientist can do big science cheaply today and in a small and easy way. Technology allows research scientists to leverage tiny budgets in astonishing ways. And each of us can now easily contribute to science as an amateur, through the growing prevalence of citizen science, in which the general public helps—often in a small, incremental way—in such tasks as data collection. From categorizing galaxies or plankton to figuring out how proteins fold, everyone can now be a part of the scientific process. And although mathematics might still be the domain of the singular genius, it, too, has a place for the hobbyist or amateur: In the mid-1990s, two high school students discovered a novel additional solution to a problem that Euclid posed and solved thousands of years ago and for which no other method had been found since. There's even an entire domain known as recreational mathematics. What these examples demonstrate is that creative experiments and the right questions are as important as ample funding and infrastructure—and that technology is making this work easier than ever. Little science can still prosper.

SADNESS IS ALWAYS BAD, HAPPINESS IS ALWAYS GOOD

JUNE GRUBER

Assistant professor of psychology; director, Positive Emotion and Psychopathology Laboratory, University of Colorado, Boulder

One idea in the study of emotion and its effect on psychological health is overdue for retirement—that negative emotions, like sadness or fear, are inherently bad or maladaptive for our psychological well-being, whereas positive emotions, like happiness or joy, are inherently good or adaptive. Such value judgments are to be understood, within the framework of affective science as depending on whether an emotion impedes or fosters a person's ability to pursue goals, attain resources, and function effectively within society. Claims such as "Sadness is inherently bad" or "Happiness is inherently good" must be abandoned in light of advances in the scientific study of human emotion.

Let's start with negative emotions. Early hedonic theories defined well-being, in part, as the relative absence of negative emotion. Empirically-based treatments like cognitive behavioral therapy also focus on the reduction of negative feelings and moods as part of enhancing well-being. Yet a strong body of scientific work suggests that negative emotions are essential to our psychological well-being. Here are three examples:

> (1) From an evolutionary perspective, negative emotions aid in our survival by providing important clues to threats or problems needing our attention, such as an unhealthy relationship or a dangerous situation.

(2) Negative emotions help us focus: They facilitate detailed and analytic thinking, reduce stereotypic thinking, enhance eyewitness memory, and promote persistence in challenging cognitive tasks.

(3) Attempting to suppress negative emotions, rather than accepting and appreciating them, paradoxically backfires; it increases feelings of distress and intensifies clinical symptoms of substance abuse, overeating, and even suicidal ideation.

Counter to these hedonic theories of well-being, negative emotions are thus not inherently bad for us. Moreover, their relative absence predicts poorer psychological adjustment.

Positive emotions are seen as pleasant or positively valenced states motivating us to pursue goal-directed behavior. A long-standing scientific tradition has focused on the benefits of positive emotions: cognitive benefits, such as enhanced creativity; social benefits, like relationship satisfaction and prosocial behavior; and physical benefits, such as enhanced cardiovascular health. From this work has emerged the assumption that positive emotional states should always be maximized, fueling the birth of entire subdisciplines and garnering much popular attention. But there's a mounting body of work against the claim that positive emotions are inherently good:

(1) They foster self-focused behavior, including selfishness, stereotyping of out-group members, cheating and dishonesty, and decreased empathic accuracy in some contexts.

(2) They're associated with greater distractibility and impaired performance on detail-oriented cognitive tasks.

(3) Because they may reduce inhibitions, they're associated with risk-taking and higher mortality rates.

Indeed, positive emotions aren't always adaptive and sometimes impede our well-being and even our survival. Valence is not value: We cannot infer value judgments about emotions based on their positive or negative valence; there's no intrinsic goodness or badness of an emotion merely because of its positivity or negativity. Instead, we must refine specific value-based determinants for an emotion's functionality. To this end, new research highlights critical variables to focus on. Importantly, the context in which an emotion unfolds can determine whether it helps or hinders an individual's goal—and which types of emotion-regulatory strategies (reappraising or distracting) will best match the situation. Moreover, the degree of one's psychological flexibility, including how quickly one can shift emotions or rebound from a stressful situation, promotes critical health outcomes.

Psychological well-being is not determined by the presence of one type of emotion but by a diversity of emotions, both positive and negative. Whether or not an emotion is "good" or "bad" seems to have surprisingly little to do with the emotion itself but rather with how mindfully we ride the ebbing and flowing tides of our rich emotional life.

OPPOSITES CAN'T BOTH BE RIGHT

ELDAR SHAFIR
William Stewart Tod Professor of Psychology and Public Affairs, Princeton University; coauthor (with Sendhil Mullainathan), Scarcity

British chef Heston Blumenthal's imaginative "Hot and Iced Tea" is a syrupy concoction prepared by putting a divider down the middle of a glass, then filling one side with a hot tea and the other with an iced version. Because of the viscous consistency of the liquid, when the divider is removed the two halves stay separate long enough for the lucky diner to simultaneously sample perfectly hot and iced tea. When you sip Blumenthal's tea, it makes no sense to argue about whether it's cold or hot. You could, of course, take care to sip only from the cold side or only from the hot. But the glass of tea is both.

Much of the world and the sciences—certainly the social and behavioral sciences—look more like that glass of tea than we often let on.

We typically assume, for example, that happiness and sadness are polar opposites and thus mutually exclusive. But recent research on emotion suggests that positive and negative affects should not be thought of as existing on opposite sides of a continuum, and that in fact feelings of happiness and sadness can co-occur. When study participants were surveyed immediately after watching certain films, or after graduating from college, they were found to feel both profoundly happy *and* sad. Our emotional experience, it turns out, is a lot like a glass of viscous tea: It can run hot and cold at the same time.

The same can be true of good and evil. Like sipping from

the hot or the cold side of the glass, we now know that minor contextual nuance can make all the difference. In one classic study, psychologists J. M. Darley and C. D. Batson recruited seminary students to deliver a sermon on the parable of the Good Samaritan. While half the seminarians were told they were comfortably ahead of the scheduled time for the talk, others were led to believe they were running late. On their way to give the talk, all participants encountered an ostensibly injured man slumped in a doorway, groaning and needing help. Most of those with time to spare stopped to help, but a mere 10 percent of those who were running late stopped, the rest stepping over the victim and rushing on.[v] Notwithstanding their ethical training and biblical scholarship, the minor nuance of a time constraint proved critical to the seminarians' decision to ignore the pleas of a suffering man. Like that high-concept glass of tea, each seminarian was *both* caring and indifferent, displaying one trait or the other depending on an arbitrary twist of fate.

Or consider John Rabe, the bald and bespectacled German engineer known as "the living Buddha of Nanking." Rabe was the legendary head of the International Committee for the Nanking Safety Zone; he was credited with having saved hundreds of thousands of Chinese lives during the savage Japanese occupation. On the other side of the glass, Rabe was simultaneously the leader of the Nazi Party in the same city. In 1938, he assured audiences that he supported the German political system "100 percent."

In its essence, this sort of anti–Manichaean perspective posits that not only one alternative always obtains. If you believe that people are only always good or always evil, if you think the glass is only either hot or cold—well, you're just wrong. You haven't felt the glass, and you have a terribly naïve understanding of human nature. But as long as your views are not that extreme, as long as you recognize the possibility of both cold and hot, then in many cases you needn't choose—it turns out they're both there.

From the little I understand, physicists question the classical distinction between wave and matter, and biologists refuse to choose between nature and nurture. But let me stay close to what I know best. In the social sciences, there's ongoing and often heated debate about whether or not people are rational and about whether or not they're selfish. And there are compelling studies in support of either camp, the hot and the iced. People can be cold, precise, selfish, and calculating. Or they can be hotheaded, confused, altruistic, emotional, and biased. In fact, they can exhibit these conflicting traits at the very same time. People can be perfectly calibrated weather forecasters but hopelessly overconfident investors, ruthless rulers and cuddly pet owners, compassionate friends and apathetic parents. Research on decisions made in demanding contexts has found that people can be thoughtful and calculating as they focus on issues of immediate concern but negligent and misguided when it comes to issues—sometimes closely related and equally (or more) important ones—at the periphery of their attention.

As we all know, history is filled with smart people who did stupid things and good people who acted horribly. Are we altruistic or selfish? Smart or stupid? Good or evil? Like that hot and iced tea, there's always a little of both—it just depends on which side of the glass you drink from.

PEOPLE ARE SHEEP

DAVID BERREBY

Journalist; blogger, Mind Matters, *bigthink.com; author,*
Us and Them: The Science of Identity

In the late summer of 1914, as European civilization began its extended suicide, dissenters were scarce. On the contrary: From every major capital, we have newsreel footage of happy crowds cheering in the summer sunshine. More war and oppression followed in subsequent decades, and there was never a shortage of willing executioners and obedient lackeys. By mid-century, the time of Stalin and Mao and their smaller-bore imitators, it seemed urgent to understand why people throughout the 20th century had failed to rise up against masters who sent them to war, or to concentration camps, or to the gulag. So social scientists came up with an answer, which was then consolidated and popularized into something every educated person supposedly knows: People are sheep—cowardly, deplorable sheep.

This idea—that most of us are unwilling to think for ourselves, instead preferring to stay out of trouble, obey the rules, and conform—was supposedly established by rigorous laboratory experiments. ("That we have found the tendency to conformity in our society so strong that reasonably intelligent and well-meaning young people are willing to call white black is a matter of concern," wrote the great psychologist Solomon Asch in 1955.)[w] Plenty of research papers still refer to one or another aspect of the sheep model as if it were a truth universally acknowledged and a sturdy rock on which to build new hypotheses about mass behavior. Worse yet, it's rampant in the conversation of educated laypeople—politicians, voters, government officials. Yet it's false. It makes for bad assumptions and bad policies. It's time to set it aside.

Some years ago, the psychologists Bert Hodges and Anne Geyer examined one of Asch's experiments from the 1950s. He'd asked people to look at a line printed on a white card and then say which of three similar lines on another card was the same length. Each volunteer was part of a small group, whose other members were actually collaborators in the study, deliberately picking wrong answers. Asch reported that when the group chose the wrong match, many individuals went along, against the evidence of their own senses.

But the experiment actually involved twelve separate comparisons for each subject, and the subjects did not agree with the majority most of the time. In fact, on average, each subject agreed three times with the majority and insisted on his own view the nine other times. To make those results all about the evils of conformity is to say, as Hodges and Geyer note, that "an individual's moral obligation in the situation is to 'call it as he sees it' without consideration of what others say."[x]

To explain their actions, the volunteer subjects didn't indicate that their senses had been warped or that they were terrified of going against consensus. Instead, they said they had chosen to go along that one time. It's not hard to see why a reasonable person would do so.

The "people are sheep" model sets us up to think in terms of obedience or defiance, dumb conformity versus solitary self-assertion (to avoid being a sheep, you must be a lone wolf). It doesn't recognize that people need to place their trust in others and win the trust of others and that this guides their behavior. (Stanley Milgram's famous experiments, in which people were willing to give severe shocks to a supposed stranger, are often cited as Exhibit A for the "people are sheep" model. But what these studies really tested was the trust the subjects had in the experimenter.)

Indeed, questions about trust in others—how it's won and kept, who wins it and who doesn't—seem essential to understanding how collectives of people operate and affect their members. What else is at work?

It appears that behavior is also susceptible to the sort of moment-by-moment influences once considered irrelevant noise (for example, seminary students in a rush were far less likely to help a stranger than were seminary students who weren't late, in the experiment performed by John M. Darley and Dan Batson). And then there's mounting evidence of influences that discomfit psychologists because there doesn't seem to be much psychology in them at all. For example, Neil Johnson of the University of Miami and Michael Spagat of Royal Holloway, University of London, and their colleagues have found the severity and timing of attacks in many different wars (different actors, different stakes, different cultures, different continents) adhere to a power law.[y] If that's true, then an individual fighter's motivation, ideology, and beliefs make much less difference than we think for the decision to attack next Tuesday.

Or, to take another example, if, as Nicholas Christakis' work suggests, your risks of smoking, getting a sexually transmitted disease, catching the flu, or being obese depend in part on your social-network ties, then how much difference does it make what you, as an individual, feel or think?

Perhaps the behavior of people in groups will eventually be explained as a combination of moment-to-moment influences (like waves on the sea) and powerful drivers that work outside of awareness (like deep ocean currents). All the open questions are important and fascinating. But they're visible only after we give up the simplistic notion that we're sheep.

BEAUTY IS IN THE EYES OF THE BEHOLDER

DAVID M. BUSS

Professor of psychology, University of Texas, Austin; author, The Dangerous Passion: Why Jealousy Is as Necessary as Love and Sex

For most of the past century, mainstream social scientists have assumed that attractiveness is superficial, arbitrary, and infinitely variable across cultures. Many still cling to these views. Their appeal has many motivations. First, beauty is undemocratically distributed, a violation of the belief that we're all created equal. Second, if physical desirability is superficial ("You can't judge a book by its cover"), its importance can be dismissed, taking a backseat to deeper and more meaningful qualities. Third, if standards of beauty are arbitrary and infinitely variable, they can be easily changed.

Two movements in the 20th century seemed to lend scientific support to these views. The first was behaviorism. If the content of human character was built through experienced contingencies of reinforcement during development, those contingencies must have created standards of attractiveness. The second was seemingly astonishing ethnographic discoveries of cross-cultural variability in attractiveness. If the Maori in New Zealand found particular types of lip tattoos attractive and the Yanomami of the Amazon rain forest prized nose or cheek piercings, then surely all other beauty standards must be similarly arbitrary.

The resurgence of sexual-selection theory in evolutionary biology, and specifically the importance of preferential mate choice, created powerful reasons to question the theoretical position long held by social scientists. We now know that in species with preferential mate choice, from scorpionflies to peacocks to elephant seals, physical appearance typically matters greatly. It conveys critical reproductively valuable qualities such as health, fertility, dominance, and "good genes." Are humans a bizarre exception to all other sexually reproducing species?

Evolutionary theorizing, long antedating the hundreds of empirical studies on the topic, suggested we were not. In mate selection, Job One, as someone in business might say, is the successful selection of a fertile partner. Those who failed to find fertile mates left no descendants. Everyone alive today is the product of a long and unbroken line of ancestors who succeeded. As evolutionary success stories, each modern human has inherited the mate preferences of his or her successful ancestors.

Cues recurrently observable to our ancestors which were reliably, statistically, probabilistically correlated with fertility, according to this theory, should become part of our evolved standards of beauty. In both genders, these include cues to health—symmetrical features and absence of sores and lesions, for example. Since fertility is sharply age-graded in women, more so than in men, cues to youth should figure prominently in gender-specific standards of attractiveness. Clear skin, full lips, an unclouded sclera, feminine estrogen-dependent features, a low waist-to-hip ratio, and many other cues to female fertility are now known to be pieces of the puzzle of universal standards of female beauty.

Women's evolved standards of male attractiveness are more complex. Masculine features, hypothesized to signal healthy immune functioning in men, are viewed as attractive more by women seeking short-term rather than long-term mates, more when women are ovulating than when they're in the luteal phase of their menstrual cycle, and more by women who are higher in mate value—perhaps because of their ability to attract and control such men. Women's judgments of men's attractiveness are more dependent on multiple contexts: cues to social status, the attention structure, positive interactions with babies, being seen with attractive women, and many others. The greater complexity and variability of what women find attractive in men is reflected in another key empirical finding: There's far less consensus among women about which men are attractive than among men about which women are attractive.

The theory that beauty is in the eye of the beholder, in the sense of being superficial, arbitrary, and infinitely culturally variable, can safely be discarded. I regard it as one of the great myths perpetrated by social scientists in the 20th century. Its scientific replacement—that beauty is in "the adaptations of the beholder," as anthropologist Donald Symons phrases it—continues to be disturbing to some. It violates some of our most cherished beliefs and values. But then so did the notion that the Earth was not flat or the center of the universe.

ROMANTIC LOVE AND ADDICTION

HELEN FISHER

Biological anthropologist, Rutgers University; author,
Why Him? Why Her? How to Find and Keep Lasting Love

"If at first the idea is not absurd, then there is no hope for it," Albert Einstein reportedly said. I'd like to broaden the definition of addiction—and also retire the scientific idea that *all* addictions are pathological and harmful.

Since the beginning of formal diagnostics more than fifty years ago, the compulsive pursuit of gambling, food, and sex (known as non-substance rewards) have not been regarded as addictions. Only abuse of alcohol, opioids, cocaine, amphetamines, cannabis, heroin, and nicotine have been formally regarded as addictions. This categorization rests largely on the fact that substances activate basic "reward pathways" in the brain associated with craving and obsession and produce pathological behaviors. Psychiatrists work within this world of psychopathology—that which is abnormal and makes you ill.

As an anthropologist, I think they're limited by this view. Scientists have now shown that food, sex, and gambling compulsions employ many of the same brain pathways activated by substance abuse. Indeed, the 2013 edition of the *Diagnostic and Statistical Manual of Mental Disorders* (the DSM) has finally acknowledged that at least one form of non-substance abuse—gambling—can be regarded as an addiction. The abuse of sex and food have not yet been included. Neither has romantic love. I shall propose that love addiction is just as real as any other addiction, in terms of its behavior patterns and brain mechanisms. Moreover, it's often a *positive* addiction.

Scientists and laymen have long regarded romantic love as part of the supernatural, or as a social invention of the troubadours in 12th-century France. Evidence does not support these notions. Love songs, poems, stories, operas, ballets, novels, myths and legends, love magic, love charms, love suicides and homicides—evidence of romantic love has now been found in more than 200 societies ranging over thousands of years. Around the world, men and women pine for love, live for love, kill for love, and die for love. Human romantic love, also known as passionate love or "being in love," is regularly regarded as a human universal.

Moreover, love-besotted men and women show all the basic symptoms of addiction. Foremost, the lover is stiletto-focused on his/her drug of choice, the love object. The lover thinks obsessively about him or her (intrusive thinking), and often compulsively calls, writes, or stays in touch. Paramount in this experience is intense motivation to win one's sweetheart, not unlike the substance abuser fixated on the drug. Impassioned lovers distort reality, change their priorities and daily habits to accommodate the beloved, experience personality changes (affect disturbance), and sometimes do inappropriate or risky things to impress this special other. Many are willing to sacrifice, even die for, "him" or "her." The lover craves emotional and physical union with the beloved (dependence). And like addicts who suffer when they can't get their drug, the lover suffers when apart from the beloved (separation anxiety). Adversity and social barriers even heighten this longing (frustration attraction).

In fact, besotted lovers express all four of the basic traits of addiction: craving, tolerance, withdrawal, and relapse. They feel a "rush" of exhilaration when they're with their beloved (intoxication). As their tolerance builds, they seek to interact with the beloved more and more (intensification). If the love object breaks off the relationship, the lover experiences signs of drug withdrawal, including protest, crying spells, lethargy, anxiety, insomnia or hypersomnia, loss of appetite or binge eating, irritability, and loneliness. Lovers, like addicts, also often go to extremes, sometimes doing degrading or physically dangerous things to win back the beloved. And lovers relapse the way drug addicts do. Long after the relationship is over, events, people, places, songs, or other external cues associated with their abandoning sweetheart can trigger memories and renewed craving.

Of the many indications that romantic love is an addiction, however, perhaps none is more convincing than the growing data from neuroscience. Using fMRI, several scientists have now shown that feelings of intense romantic love engage regions of the brain's "reward system": specifically, dopamine pathways associated with energy, focus, motivation, ecstasy, despair, and craving, including primary regions associated with substance (and non-substance) addictions. In fact, I and my colleagues Lucy Brown, Art Aron, and Bianca Acevedo have found activity in the nucleus accumbens—the core brain factory associated with all addictions—in rejected lovers. Moreover, some of our newest results suggest correlations between activities of the nucleus accumbens and feelings of romantic passion among lovers who are wildly, happily in love.

Nobel laureate Eric Kandel has noted that brain studies "will give us new insights into who we are as human beings."[z] Knowing what we now know about the brain, my brain-scanning partner Lucy Brown has suggested that romantic love is a natural addiction, and I've maintained that this natural addiction evolved from mammalian antecedents some 4.4 million years ago among our first hominid ancestors, in conjunction with the evolution of (serial, social) monogamy—a hallmark of humankind. Its purpose: to motivate our forebears to focus their mating time and metabolic energy on a single partner at a time, thus initiating the formation of a pair-bond to rear their young (at least through infancy) together as a team.

The sooner we embrace what brain science is telling us—and use this information to upgrade the concept of addiction—the better we'll understand ourselves and the billions of others on this planet who revel in the ecstasy and struggle with the sorrow of this profoundly powerful, natural, often positive addiction: romantic love.

EMOTION IS PERIPHERAL

BRIAN KNUTSON

Associate professor of psychology and neuroscience, Stanford University

Some still assume that emotion is peripheral, but the time has come to recognize that emotion is central.

The claim that emotion is peripheral can be taken both literally and figuratively. From a literal standpoint, experts have argued since the birth of experimental psychology in the Gilded Age about which physiology is necessary for emotional experience. In his seminal essay "What Is an Emotion?" William James counterintuitively claimed that when we encounter a bear, peripheral (i.e., below the neck) physiological changes occur (the stomach clenches, the heart pounds, the skin sweats), which then generate an experience of emotion (fear). By implication, peripheral responses must occur before the feeling of fear.

His Harvard colleague Walter Cannon disagreed, stating that brain activity causes both emotional experience and peripheral responses. Cannon based his argument on research (for example, emotional responses could be evoked by stimulating the brains of cats, who continued to show those emotional responses after spinal cord lesions) as well as on physiological logic (peripheral responses were too slow, insensitive, and undifferentiated to drive emotional experience). Thus, although James was a creative thinker and a persuasive writer, he reasoned from the armchair, whereas the stolid and understated Cannon (who also innovated influential concepts such as "homeostasis" and "fight or flight") brought data to bear on the debate.

I keep revisiting this century-old academic scuffle. That's because peripheralist assumptions still form the backbone of many modern emotion theories (e.g., in the form of peripheral somatic signals, or embodiment, or indeed any sensory process purported to mediate emotion). Of course, peripheral responses can modulate emotion, but they're not fast or specific enough to mediate the kinds of rapid emotional responses that ensured our ancestors' survival. Emotion also undoubtedly generates peripheral responses, but without information about which came first, correlated action doesn't imply causal direction. To be fair to the peripheral view, scientists currently lack a quantitative computational model of exactly how the brain generates emotion, and the neural mechanisms are still being worked out. But as the next few years of brain-stimulation, lesion, and imaging evidence accumulates, I'm betting that the central account of emotion will prevail.

From a figurative standpoint, the problematic assumptions of emotional peripheralism run deeper. An even older debate focuses on emotion's function rather than structure. Specifically, is emotion peripheral or central to mental function? A peripheralist viewpoint might posit that emotion does not influence, and even disrupts, mental function. While the historical roots of such an assumption may reach back as far as Zoroastrian dualism, René Descartes typically gets the blame for importing dualism from the church to science. Descartes split the mind and passions by placing the mind with the spirit but the passions with the body (where they took the form of "animal spirits" purported to move the pineal gland). According to Cartesian mind-body dualism, the mind could thus operate independently of the disruption of excessive passions.

In contrast to this peripheralist vision, a depiction of the centrality of emotion to mental function comes not from the West but from the East. The Tibetan Buddhist "Wheel of Life" represents passionate attachments as animals occupying the hub of a spinning wheel, driving thought and behavior. In both schemes, excessive passions can divert thought and action, but in Descartes' scheme, emotion disrupts the mind from the periphery, whereas in the Buddhist scheme emotion drives the mind from the center. If emotion is central to mental function, then our inherited scientific map of the mind is inside-out.

Indeed, the absence of emotion pervades modern scientific models of the mind. In the most popular mental metaphors of social science, mind as reflex (from behaviorism) explicitly omits emotion, and mind as computer (from cognitivism) all but ignores it. Even when emotion appears in later theories, it's usually an afterthought—an epiphenomenal reaction to something that has already happened. But over the past decade, the rising field of affective science has revealed that emotions can precede and motivate thought and behavior.

Emerging physiological, behavioral, and neuroimaging evidence suggests that emotions are proactive as well as reactive. Emotional signals from the brain now yield predictions about choice and mental-health symptoms and may soon guide scientists to specific circuits that confer more precise control over thought and behavior. Thus, the price of continuing to ignore emotion's centrality to mental function could be substantial. By assuming the mind is like a bundle of reflexes, a computer program, or even a self-interested rational actor, we may miss out on significant opportunities to predict and control behavior—both in individuals and groups. We should stop relegating emotion to the periphery and move it to the center, where it belongs.

SCIENCE CAN MAXIMIZE OUR HAPPINESS

PAUL BLOOM

Brooks and Suzanne Ragen Professor of Psychology and Cognitive Science, Yale University; author, Just Babies: The Origins of Good and Evil

Psychologists have made striking discoveries about what makes people happy. Some of these findings clash with common sense. It turns out, for instance, that we're much better than we think we are at rebounding from negative experiences—we're usually blind to the workings of what Harvard psychologist Daniel Gilbert calls our "psychological immune system." Other discoveries mesh with what our grandmothers could have told us, such as the happiness boost from being with friends and the misery that often comes from solitude. Better to live as Donald Duck than Scrooge McDuck.

Some leading researchers believe that as this work proceeds, we'll converge on a scientific solution as to how to maximize our happiness. This is mistaken. Even assuming a perfectly objective definition of happiness (and putting aside the distinction between a happy life and a *good* life), the issue of how to construct a maximally happy life falls, at least in part, outside the domain of science.

To see why, consider a related question: How can we determine the happiest society? As the British philosopher Derek Parfit and others have pointed out, even if you can precisely measure the happiness of each individual, this remains a vexingly hard question. Should we choose the society with the highest total happiness? If so, then a trillion people living miserable lives (but not so miserable that they'd rather be dead) will be "happier" than a billion immensely happy people.

This seems wrong. Do we calculate averages? If so, then a society with a majority of extremely happy individuals and a small minority who suffer terrible torment might be "happier" than a society where everyone is merely very happy. This seems wrong, too. Or consider the contrast between (a) a society in which people are equally happy versus (b) a society with gross inequality but which has both a larger total happiness and a larger average happiness than (a). Which is happier? This is a hard problem, with real-world relevance, and it isn't the sort of problem that will be solved by science, because science provides no empirical recipe for how overall happiness should be calculated.

Importantly, as Parfit notes, the same problems arise with regard to an individual life. How should you balance your happiness across a lifetime? Which life is happier—one that's somewhat happy throughout or one that's a balance between joy and misery? Again, this isn't the sort of question that can be answered experimentally.

Then there are moral concerns. We're often faced with situations in which we have to choose whether to sacrifice our own happiness for the benefit of others. Most of us make such sacrifices for friends and families; some of us do so for strangers. Framed this way, it's a moral problem, not a hedonic one: A perfect hedonist would help others only to the extent that she believed it would increase her own happiness. But now consider that the same tradeoffs apply for a single individual, within a single life span. Think of your happiness now, and ask yourself how much you'd give up—not for another person but for yourself in the future.

Life is full of such choices. When we indulge in certain immediate pleasures—fatty foods, unsafe sex, living like there's no tomorrow—we're greedily maxing out on our happiness now at the expense of the happiness of our future selves. When we sacrifice for the future—unpleasant exercise, healthy and tasteless foods, saving for a rainy day—we're altruists, sacrificing now for the happiness of our future selves. Surprisingly, then, even the most selfish hedonist has to wrestle with moral questions, and seeming scientific questions about happiness quickly turn into manifestly nonscientific questions about the right thing to do.

CULTURE

PASCAL BOYER

Anthropologist, psychologist, Henry Luce Professor of Individual and Collective Memory, Washington University, St. Louis; author, Religion Explained

Culture is like trees. Yes, there are trees around. But that doesn't mean we can have a science of trees. Having some rough notion of "tree" is useful for snakes that lurk and fall on their prey, for birds that build nests, for humans trying to escape from rabid dogs, and of course for landscape designers. But the notion is of no use to scientists. There's nothing much to find out—for example, to explain growth, reproduction, evolution—that would apply to all and only those things humans and snakes and birds think of as "trees." Nothing much that would apply to both pines and oaks, to both baobabs and monstrous herbs like the banana tree.

Why do we think there's such a thing as culture? Like "tree," it's a convenient term. We use it to designate all sorts of things we feel need a general term, like the enormous amount of information that humans acquire from other humans, or the set of idiosyncratic concepts or norms we find in some human groups but not in others. There's no evidence that either of these domains corresponds to a proper set of things that science could study and about which it could offer general hypotheses or describe mechanisms.

Don't get me wrong. We can and should engage in a scientific study of "cultural stuff." Against the weird obscurantism of many traditional sociologists, historians, or anthropologists, human behavior and communication can and should be studied in terms of their natural causes. But this doesn't imply that there will or should be a science of culture in general.

We can run scientific studies of general principles of human behavior and communication—that's what evolutionary biology and psychology and neuroscience can do—but that's a much broader domain than "culture." Conversely, we can run scientific studies of such domains as the transmission of technologies, or the persistence of coordination norms, or the stability of etiquette—but these are much narrower domains than "culture." About cultural stuff, as such, in general, I doubt any good science can say anything.

This in a way is not surprising. When we say that some notion or behavior is "cultural," we're just saying that it bears some similarity to notions and behaviors of other people. That's a statistical fact. It doesn't tell us much about the processes causing that behavior or notion. As the French cognitive scientist Dan Sperber put it, cultures are epidemics of mental representations. But knowing the epidemiological facts—that this idea is common whereas that one is rare—is of no use unless you know the physiology, so to speak: how this idea was acquired, stored, modified, how it connects to other representations and to behavior. We can say lots of interesting things about the dynamics of transmission, and scholars from Rob Boyd and Pete Richerson to more recent modelers have done just that. But such models don't aim to explain why cultural stuff is the way it is—and there's probably no general answer to that.

Is the idea of culture really a Bad Thing? Yes, a belief in culture as a domain of phenomena has hindered the development of a proper science of human behavior in groups—what ought to be the domain of social sciences.

First, if you believe there's such a thing as "culture," you naturally tend to think it's a special domain of reality with its own laws. But it turns out that you cannot find the unifying causal principles (because there aren't any). So you marvel at the many-splendored variety and diversity of culture. But culture is splendidly diverse only because it's not a domain at all, just as there's a marvelous variety in the domain of white objects or of people younger than Socrates.

Second, if you believe in culture as a thing, it seems normal to you that culture should be the same across individuals and across generations. So you treat as unproblematic precisely the phenomenon that's vastly improbable and deserves a special explanation. Human communication doesn't proceed by direct transfer of mental representations from one brain to another. It consists in inferences from other people's behaviors and utterances, which rarely if ever leads to the replication of ideas. That such processes could lead to roughly stable representations across large numbers of people is a wonderful anti-entropic process that cries out for explanation.

Third, if you believe in culture, you end up believing in magic. You'll say that some people behave in a particular way because of "Chinese culture" or "Muslim culture." In other words, you'll be trying to explain material phenomena—representations and behaviors—in terms of a nonmaterial entity, a statistical fact about similarity. But a similarity doesn't cause anything. What causes behaviors are mental states.

Some of us aim to contribute to a natural science of human beings as they interact and form groups. We have no need for that social-scientific equivalent of phlogiston, the notion of culture.

CULTURE

LAURA BETZIG

Anthropologist, historian; author, Despotism and Differential Reproduction

Years ago, when I sat at the feet of the master, the King of the Amazon Jungle liked to talk about culture. He quoted his own teachers, who considered it *sui generis:* Culture was a thing in and of itself. It made us more than the sum of our biological parts; it emancipated us from the Promethean bonds of our evolutionary past. It set us apart from other animals and made us special.

Napoleon Chagnon wasn't so sure about that, and neither was I.

What if the 100,000-year-old evidence of human social life—from arrowheads in South Africa to Venus figurines at Dordogne—is the effect of nothing more or less but our efforts to become parents? What if the 10,000-year-old record of civilization—from tax accounts at Near Eastern temples to the inscription on a bronze statue in New York Harbor—is the product of nothing more or less but our struggle for genetic representation in future generations?

Either case can be made. For 100,000 years or more, prehistoric foragers probably lived like contemporary foragers in Africa or Amazonia. They probably did their best to live in peace but occasionally fought over the means of production and reproduction, so that the winners cohabited with more women and supported more children. And they probably were more likely to fight where it was harder to flee, in territory where resources were easy to come by and food and shelter in nearby territories were relatively scarce.

Then, within just the last 10,000 years, the first civilizations were built. From Mesopotamia to Egypt, from India to China, then in Greece and Rome, eusocial emperors—like eusocial insects—turned some of their subordinates into sterile castes but were extraordinarily fertile themselves. A *praepositus sacri cubiculi,* or eunuch, set over the sacred bedchamber, eventually ran the empire on the Tiber, and other eunuchs collected revenues, led armies, and kept track of the hundreds of "homeborn" children in the *Familia Caesaris*—the imperial family in Rome. Then the barbarians invaded, and the emperor took his slave harem off to a secure spot on the Bosporus.

And the Republic of St. Peter took over in the depopulated west. From Clovis's kingdom in Paris to Charlemagne's empire at Aachen to the Holy Roman conglomerate east of the Rhine, cooperatively breeding aristocrats—like cooperatively breeding birds—turned some of their sons and daughters into celibates but raised others to become husbands and wives. Abbesses, abbots, and bishops administered estates and conscripted troops, or instructed their nieces and nephews in monastery schools; their older brothers begot heirs to their enormous castles or covered the countryside with bastards. Then the Crusaders took ships to the Near East, and Columbus led the first waves of immigrants across the Atlantic.

Over the next few centuries, hordes of poor, huddled masses from across the Old World found places to breathe free on the American continents. Millions of solitary slaves and serfs, and thousands of unmarried priests and monks—like helper birds or social-insect workers whose habitats had opened up—walked away from their lords and masters and out of their cathedrals and abbeys. They were hoping to secure liberty for themselves and their posterity; they were looking for places to raise their own families. In the *Common Sense* words of a common man, Tom Paine: "Freedom hath been hunted round the globe, Asia, and Africa have long expelled her. Europe regards her like a stranger, and England hath given her warning to depart. O! receive the fugitive and prepare in time an asylum for mankind."

Since those early days when I learned from Napoleon Chagnon, it has seemed to me that "culture" is a seven-letter word for God. Good people (some of the best) and intelligent people (some of the smartest) have found meaning in religion: They have faith that something supernatural guides what we do. Other good, intelligent people have found meaning in culture: They believe that something superzoological shapes the course of human events. Their voices are often beautiful, and it's wonderful to be part of a chorus. But in the end, I don't get it. For me, the laws that apply to animals apply to us.

And in that view of life, there is grandeur enough.

LEARNING AND CULTURE

JOHN TOOBY

Founder of evolutionary psychology; codirector, Center for Evolutionary Psychology; professor of anthropology, University of California, Santa Barbara

Any firsthand experience of how scientific institutions actually operate drives home an excruciating realization: Science progresses more slowly by orders of magnitude than it could or should. Our species could have science at the speed of thought—science at the speed of inference. But too often we run into Planck's demographic limit on the speed of science—funeral by funeral, with each tock of advancement clocked to the half-century tick of gatekeepers' professional life spans.

In contrast, the natural clock rate of science at the speed of thought is the flash rate at which individual minds, voluntarily woven into mutually invigorating communities by intense curiosity, can draw and share sequences of strong inferences from data. Indeed, Planck was a giddy optimist, because scientists—like other humans—form coalitional group identities where adherence to group-celebrating beliefs (e.g., "We have it basically right") are strongly moralized.

So, the choice is frequently between being "moral" or thinking clearly. Because the bearers of reigning orthodoxies educate and self-select their next-generation replacements, mistakes not only propagate down generations but can grow to Grand Canyon size. When this happens, data sets become embedded so deeply into a matrix of mistaken interpretations (as in the human sciences) that they can no longer be seen independently of their obscuring frameworks. So the sociological speed of science can end up being slower even than Planck's glacial demographic speed.

Worst of all, the flow of discoveries and better theories through institutional choke points is clogged by ideas that are so muddled that they are—in Wolfgang Pauli's telling phrase—not even wrong. Two of the worst offenders are learning and its partner in crime, culture—a pair of deeply established, infectiously misleading, yet (seemingly) self-evidently true theories. What alternative to them could there be, except an easily falsified robotic genetic determinism?

Yet countless obviously true scientific beliefs have had to be discarded—a stationary Earth, (absolute) space, the solidity of objects, no action at a distance, and so on. Like these others, learning and culture seem compelling because they map closely to automatic, built-in features of how our minds evolved to interpret the world (for example, learning is a built-in concept in the theory-of-mind system). But learning and culture are not scientific explanations for anything. Instead, they're phenomena that themselves require explanation.

All "learning" operationally means is that something about the organism's interaction with the environment caused a change in the information states of the brain, by mechanisms unexplained. All "culture" means is that some information states in one person's brain somehow cause, by mechanisms unexplained, "similar" information states to be reconstructed in another's brain. The assumption is that because supposed instances of "culture" (or, equally, "learning") are referred to with the same name, they're the same kind of thing. Instead, each masks an enormous array of thoroughly dissimilar things. Attempting to construct a science built around culture (or learning) as a unitary concept is as misguided as attempting to develop a robust science of white things (eggshells, clouds, O-type stars, Pat Boone, human scleras, bones, first-generation MacBooks, dandelion sap, lilies . . .).

Consider buildings and the things that allow them to influence one another: roads, power lines, water lines, sewage lines, mail, phone landlines, sound, wireless phone service, cable, insect vectors, cats, rodents, termites, dog-to-dog barking, fire spread, odors, line-of-sight communication with neighbors, cars and delivery trucks, trash service, door-to-door salesmen, heating-oil delivery, and so on. A science whose core concept was building-to-building influence ("building culture") would be largely gibberish, just as our "science" of culture as person-to-person influence has turned out to be.

Culture is the functional equivalent of protoplasm, the supposed (and "observed") substance that by mechanisms unknown carried out vital processes. Now we recognize that protoplasm was magician's misdirection—a black-box placeholder for ignorance, eclipsing the bilipid layers, ribosomes, Golgi bodies, proteasomes, mitochondria, centrosomes, cilia, vesicles, spliceosomes, vacuoles, microtubules, lamellipodia, cisternae, etc. that were actually carrying out cellular processes.

Like protoplasm, culture and learning are black boxes, imputed to have impossible properties and masquerading as explanations. They need to be replaced with maps of the diverse cognitive and motivational "organelles" (neural programs) that actually do the work now attributed to learning and culture. They're the La Brea tar pits of the social and behavioral sciences. After a century of wrong turns, our scientific vehicles continue to sink ever deeper into these tar pits, and yet we celebrate, because these conceptual tars have poured in to fill all explanatory gaps in the human sciences. They unfalsifiably "solve" all apparent problems by stickily obscuring the actual causal specificity that in each case needs to be discovered and mapped.

We overattribute our mental content to culture because the sole supposed alternative is genes. Instead, evolved, self-extracting AI-like expert systems, in interaction with environmental inputs, neurally develop to populate our minds with immense, subtle bodies of content, only some of which are sourced from others. Rather than humans as passive receptacles haplessly filled by "culture," these self-extracting systems make humans active agents robustly building their worlds. Some neural programs, in order to better carry out their particular functions, evolved to supplement their own self-generated content with low-cost, useful information drawn from others ("culture").

But like buildings, humans are linked with many causally distinct pathways built to perform distinct functions. Each brain is bristling with many independent "tubes" that propagate many distinct kinds of stuff to and from a diversity of brain mechanisms in others. So there's fear-of-snakes culture (living "inside" the snake-phobia system), grammar culture (living "inside" the language-acquisition device), food-preference culture, group-identity culture, disgust culture, sharing culture, aggression culture, and so on.

Radically different kinds of "culture" live inside distinct computational habitats—that is, habitats built out of different evolved mental programs and their combinations. What really ties humans together is an encompassing metaculture—our species' universal cognitive and emotion programs and the implicit (and hence invisible) universally shared world of meaning they give rise to. Because the adaptive logics of these evolved neural programs can now be mapped, the prospect of a rigorous natural science of humans is open to us. If we could pension off learning and culture, that would remove two obstacles to the human sciences advancing at the speed of thought.

"OUR" INTUITIONS

STEPHEN STICH
Board of Governors Professor, Department of Philosophy and Center for Cognitive Science, Rutgers University

There's a strategy for defending philosophical views which has been around since antiquity. It's used to support rules for reasoning (in science and elsewhere) and moral principles, and to defend accounts of phenomena like knowledge, causation, and meaning. Recent findings increasingly demonstrate that after 2,500 years it's a strategy ready for retirement.

Here's how it works. A case—sometimes real, often imaginary—is described, and the philosopher asks, "What would we say about that case? Does the protagonist in the story really have knowledge? Is the behavior of the protagonist morally permissible? Did the first event cause the second?" When things go well, the philosopher and his audience will make the same spontaneous judgment about the case.

Contemporary philosophers call those judgments "intuitions." And in philosophical theorizing, our intuitions are an important source of evidence. If a philosopher's theory comports with our intuition, the theory is supported; if the theory entails the opposite judgment, the theory is challenged. If you've ever taken a philosophy course, you'll likely find this method familiar. But it's not just a method that philosophers use in the classroom. At a recent colloquium in my department, I sat in the back and counted the appeals to our intuition made by a rising star in the philosophical profession during a fifty-five-minute talk. There were twenty-six—roughly one every two minutes.

That's a lot of intuition-mongering, although it's hardly unusual in contemporary philosophy. Another thing about the talk that wasn't at all unusual was that the speaker never once told us who "we" are. When a philosopher makes claims about "our" intuitions about knowledge or causation or moral permissibility, whose intuitions is he talking about? Until recently, philosophers almost never confronted that question. But if they had, their answer would likely have been inclusive. The intuitions we use as evidence in philosophy are the intuitions that all rational people would have, provided they were paying attention and had a clear understanding of the case evoking the intuition. According to contemporary defenders of this methodology, intuitions are rather like perceptions—they're shared by just about everyone.

Some of us have long thought that there was room for a fair amount of skepticism here. How could philosophers, seated comfortably in their armchairs, be so confident that all rational people share *their* intuitions? This skepticism was reinforced with the emergence of cultural psychology over the last three decades. Culture, it turns out, runs deep, and it affects a wide array of psychological processes, ranging from reasoning to memory to perception.

Moreover, in an important article, Joseph Henrich, Steven J. Heine, and Ara Norenzayan have made a persuasive case that WEIRD people (people in cultures that are Western, Educated, Industrialized, Rich, and Democratic) are outliers on a wide range of psychological tasks. WEIRD people, they argue, are "the weirdest people in the world."[aa] And philosophers are overwhelmingly WEIRD. They are also overwhelmingly white, predominantly male, and have all survived years of undergraduate and graduate training in settings where people who don't share the professionally favored intuitions are sometimes at a considerable disadvantage. Could it be that these factors, singly or in combination, explain the fact that professional philosophers, and their successful students, share lots of intuitions?

About a decade ago, this question led a group of philosophers, along with sympathetic colleagues in psychology and anthropology, to stop assuming that their intuitions were widely shared and design studies to see if they really are. In study after study, it turned out that philosophical intuitions do indeed vary with culture and other demographic variables. A great deal more work will be needed before we have definitive answers about which philosophical intuitions vary and which, if any, are universal.

There are lots of important intuitions to look at, lots of cultural and demographic groups to consider, and lots of methodological pitfalls to discover and avoid. But not surprisingly, the early efforts of these "experimental philosophers" have not been warmly welcomed by philosophers deeply invested in the traditional intuition-based method. One leading philosopher proclaimed that experimental philosophers "hate philosophy." He and others have also staked out a fallback position that insists it doesn't much matter what we discover about the intuitions of ordinary people, or of people in other cultures, because professional philosophers are the experts in making judgments about knowledge, morality, causation, and the rest, so only *their* intuitions are to be taken seriously.

It will be a long time before the dust settles in this dispute. But one conclusion on which perhaps most of those involved can agree is that it's time to stop talking about "our" intuitions without bothering to say who "we" are.

WE'RE STONE AGE THINKERS

ALUN ANDERSON

Senior consultant, former editor-in-chief and publishing director, New Scientist*; author,* After the Ice: Life, Death, and Geopolitics in the New Arctic

Back in the 1970s, the Nobel–prize winning ethologist Niko Tinbergen liked to trace out a graph. One line on it rose slowly over time, showing the rate of our genetic evolution; a second curved steeply upward, showing the rate at which he saw our culture changing. He would speculate as to whether the gap between the environment we'd evolved in and the one in which we now found ourselves might be the root of a number of ills. Since then, such ideas have spread—in part because of the rise of evolutionary psychology.

In its strong form, evolutionary psychology holds that the human mind is like a Swiss Army knife, made up of many innate special-purpose modules, each shaped by natural selection to solve problems encountered during *Homo's* long pre-civilization life. With 99 percent of our evolutionary past spent as hunter-gatherers, it seems reasonable to conclude that modules that were adaptive in past circumstances still dominate our thinking. Thus women will naturally find athletic men—the kind who would be good hunters—especially attractive. If we had instead spent the Pleistocene delving the earth like Tolkien's dwarves, then short barrel-chested men would now appeal. In the popular imagination, evolutionary psychology has cast us as Stone Age thinkers in modern times, our brains not wired to cope with offices, schools, courts, writing, or new technology.

It's a beguiling idea, suggesting that somewhere out there is a more natural world in which we'd truly feel at home. But there's little evidence for it—or that the whole of our psychology is shaped so rigidly by our Pleistocene past. It's time to retire the idea and think more widely.

New ideas and data from the cognitive sciences, comparative animal behavior, and evolutionary developmental biology suggest that we should not compartmentalize culture and human nature so sharply. Culture and social processes shape our brains, which in turn shape our culture and transmit it onward.

Reading provides a nice example. The ability to pass on and accumulate information has transformed our world, but written languages appeared only 5,000 years ago—not long enough for us to have evolved an innate "reading module." Still, if you look inside the brain of a literate person, it will light up differently from the brain of an illiterate, not just when he's reading but also when he's listening to spoken words. In the social process of being taught to read, infant brains are remodeled and new pathways created. If we didn't know that this cognitive capacity was produced by social learning, we'd likely think of it as a genetically inherited system. But it isn't. Our brains and minds can be transformed through the acquisition of cognitive tools—tools we can pass on, again and again.

Of course, it's reasonable to assume that those cognitive tools have to fit nicely with how our brain works, just as a physical tool has to fit well in our hands. But as a species we seem to possess a remarkable ability to keep building and rebuilding our cognitive toolkit through interaction with others. It's surprising how similar humans and chimpanzees are when they're infants—in skills like numeracy and behavior reading—yet how different they are as adults. Beyond a certain age, humans are propelled along a different developmental trajectory, in part because they're socially motivated to interact with others, which chimpanzees are not. Evolutionary developmental psychology has thus become a hot research topic, holding the key to the way social processes unfold minds.

Culture and the social world shape our brains and give us new cognitive capacities we can pass along, evolving culture as we go. We should think of the cultural world not as separate and estranged from our biological selves but as something that shapes us and is in turn transmitted by us. Such a view suggests that rather than being alienated hunter-gatherers lost in the modern world, we are in flux and still may have only a narrow conception of what humans could be.

INCLUSIVE FITNESS

MARTIN NOWAK

Professor of biology and mathematics; director, Program for Evolutionary Dynamics, Harvard University; coauthor (with Roger Highfield), SuperCooperators

This year marks the fiftieth anniversary of the introduction of inclusive fitness, the highly influential idea that supposedly explains how insects evolve complex societies and how natural selection can lead to altruism among relatives. This mainstay of sociobiology is based on the 1964 work of the English evolutionary biologist William Hamilton, who coined the following definition:

> Inclusive fitness may be imagined as the personal fitness which an individual actually expresses in its production of adult offspring as it becomes after it has been first stripped and then augmented in a certain way. It is stripped of all components which can be considered as due to the individual's social environment, leaving the fitness which he would express if not exposed to any of the harms or benefits of that environment. This quantity is then augmented by certain fractions of the quantities of harm and benefit which the individual himself causes to the fitnesses of his neighbours. The fractions in question are simply the coefficients of relationship appropriate to the neighbours whom he affects: unity for clonal individuals, one-half for sibs, one-quarter for half-sibs, one-eighth for cousins, . . . and finally zero

for all neighbours whose relationship can be considered negligibly small.[ab]

Modern formulations of inclusive-fitness theory use different relatedness coefficients, but all other aspects of Hamilton's definition remain intact.

Leaving aside the inelegance of Hamilton's original formulation, there's a basic problem with inclusive fitness: You can prove mathematically that inclusive fitness does not apply to the vast majority of evolutionary processes. The reason is simple. Fitness effects cannot in general be written as the sum of components caused by pairwise interactions. This loss of additivity typically occurs when the outcome of a social interaction depends on the strategies of more than one individual. All mathematically meaningful approaches to inclusive fitness realize these limitations. Thus, inclusive fitness becomes a very particular way to calculate evolution: It works in some cases but not in general. Moreover, if an inclusive-fitness calculation can be performed, it gives the same answer as a standard calculation of fitness and natural selection. The latter approach is usually simple and direct.

These mathematical facts make uncomfortable reading for overly enthusiastic proponents of inclusive fitness. In the most extreme cases, they come across as followers of a cult who believe that inclusive fitness is an important extension of the theory of evolution and "always true." In order to maintain the idea that inclusive fitness can always be calculated, a method has been devised that casts any evolutionary change in terms of virtual cost-and-benefit parameters, which appear as regression coefficients in a statistical analysis. The problem with adopting this statistical approach is that the resulting cost-

and-benefit parameters are meaningless quantities, in the sense that they don't explain what's going on in a theoretical model or in empirical data.

Why do we have inclusive fitness? Hamilton's original goal was to find a quantity that's maximized by evolution. This view is attractive; winners of the evolutionary process should be individuals with the highest inclusive fitness. But such an attempt is very much in the spirit of the linear thinking of the 1960s, before the likes of Oxford zoologist Robert May showed us how nonlinear phenomena apply to ecology, population genetics, and evolutionary game theory. From the 1970s onward, we understood that evolution does not permit a single quantity that is always maximized. This fact still has to sink in with many in the inclusive-fitness community.

What shall we use instead of inclusive fitness? Inclusive fitness seeks to explain social evolution on the level of the individual. For most evolutionary processes, however, the individual is the wrong unit of analysis, because the population structure is complicated and the same genes are present in different types of individuals. Therefore, we have to go to the level of genes. A straightforward approach is to calculate how natural selection changes the frequency of genetic mutations that affect social behavior. These calculations, which don't use inclusive fitness, can identify the key parameters that need to be measured to improve understanding. On the level of genes, there's no inclusive fitness.

We have a strong and meaningful mathematical theory of evolution. Natural selection, mutation, and population structure can all be investigated with mathematical formalism. Everyone who understands the mathematical theory of evolution realizes there's no problem that would require the calculation of inclusive fitness. Calculating inclusive fitness is an optional exercise, one best done when a problem is already completely understood. Then, in some cases, inclusive fitness can be used to re-derive the same result.

To be fair, inclusive fitness has stimulated much empirical and theoretical work over the years, some of which has been useful. It has induced a discussion of cost, benefit, and relatedness in sociobiology, which has some merit. But the dominant and unfortunate effect has been the suppression of meaningful mathematical theories in wide areas of sociobiology.

Contrary to what's often claimed, there exists no empirical test of inclusive fitness theory; nobody has ever performed an actual inclusive fitness calculation for a real population. Inclusive fitness was originally understood as a crude heuristic that could guide intuition in some cases but not in general. Only in recent years has inclusive fitness been elevated (mostly by mediocre theoreticians) to a religious belief that is universal, unconstrained, and always true. Understanding its limitations gives us the opportunity to develop mathematical descriptions of key phenomena in social evolution. It's time to abandon inclusive fitness and focus on a meaningful interaction between theory and experiment in sociobiology.

HUMAN EVOLUTIONARY EXCEPTIONALISM

MICHAEL McCULLOUGH

Director, Evolution and Human Behavior Laboratory, professor of psychology, University of Miami; author, Beyond Revenge: The Evolution of the Forgiveness Instinct

Humans are biologically exceptional. We're exceptionally long-lived and exceptionally cooperative with non-kin. We have exceptionally small guts and exceptionally large brains. We have an exceptional communication system and an exceptional ability to learn from other members of our species. Scientists love to study biologically exceptional human traits such as these, and that's a perfectly reasonable research strategy. Human *evolutionary* exceptionalism, however—the tendency to assume that biologically exceptional human traits come into the world through exceptional processes of biological evolution—is a bad habit we need to break. Human evolutionary exceptionalism has sown misunderstanding in every area it has touched. Here are three examples:

Human niche construction. Humans have exerted biologically exceptional effects on their environments. In our evolutionary past, these so-called niche-construction effects occasionally created the necessary and sufficient condition for natural selection: a generationally persistent covariance between genes and fitness. For example, earlier hominins' experimentation with cooking (which required the generationally persistent availability of culturally transmitted knowledge on how to control fire) made their food more digestible. Consequently, genetic mutations that shrank the human gut, teeth, and jaw muscles were naturally selected, because they enabled resources to be reassigned to the construction of new adaptive faculties (including cognitive ones).

For years, niche-construction theorists have argued that standard evolutionary theory cannot account for such interactions between humans' culturally mediated environmental effects and natural selection. In response, they have promoted niche construction as a "neglected evolutionary process" that collaborates with natural selection to direct evolution. However, they obtain persuasive force for this argument by redefining what evolution is. Humans' niche-construction activities have undoubtedly exposed new covariances between genetic variation and fitness during human evolution, but those activities have neither created that variation nor filtered it, so they don't constitute an evolutionary process. Culturally mediated human niche construction is real, important, sometimes evolutionarily significant, and certainly worthy of study, but it doesn't compel a revision of our understanding of how evolution works.

Major Evolutionary Transitions. Over the past 3 billion years, natural selection has yielded several pivotal innovations in how genetic information gets assembled, packaged, and transmitted across generations. These so-called major evolutionary transitions have included the transition from RNA to DNA; the union of genes into chromosomes; the evolution of eukaryotic cells; the advent of sexual reproduction; the evolution of multicellular organisms; and the appearance of eusociality (notably, among ants, bees, and wasps) in which only a few individuals reproduce and the others work as servants, soldiers, or babysitters. The concept of major evolutionary transitions, when properly applied, is useful and clarifying.

It's therefore regrettable that the concept's originators made category mistakes by characterizing two distinctly human traits as outcomes of major evolutionary transitions. Their first category mistake was to liken human societies (which are exceptional among the primates for their nested levels of organization, their mating systems, and 100 other features) to those of the eusocial insects because the individuals in both kinds of societies "can survive and transmit genes . . . only as part of a social group."[ac] This is an unfortunate case of science by analogy: The fact that humans are adapted for living in social groups doesn't imply that they—like ants, bees, wasps, and termites—need groups to reproduce. If the chemistry, timing, and lighting are just right, any human male and any human female, plucked from their social groups at random, can manage to convey genetic information to the next generation just fine.

Their second category mistake was to hold up human language as the outcome of major evolutionary transition. To be sure, human language, as the only communication system with unlimited expressive potential that natural selection ever devised, is indeed biologically exceptional. However, the information that language conveys is contained in our minds, not in our chromosomes. We don't yet know precisely where or when human language evolved, but we can be reasonably confident about *how* it evolved: via the gene-by-gene design process called natural selection. No major evolutionary transition was involved.

Human Cooperation. Humans are exceptionally generous, particularly toward nonrelatives. We cooperate with strangers when we'd be better off in the short term by competing. We donate anonymously to charities. We accomplish group projects even though all participants surely recognize that they'd be better off, at least in the short term, by loafing and letting others do the work. We share with needy strangers even when we know they'll never repay us. We praise generosity and denounce stinginess, even when the behaviors in question haven't affected us directly.

These cooperation-related phenomena were once on evolutionary scientists' lists of unsolved puzzles about human cooperation. The good news is that scientists have now succeeded in nudging many of them toward the solved-puzzles list. The bad news is that some scholars have gone in the opposite direction: They have moved these problems onto the list of "mysteries"—problems so perplexing that we should abandon hope of ever solving them within the standard inclusive-fitness-maximizing view of natural selection. Their mystification has led them, at turns, to invoke evolutionary explanations inappropriate for species in which all individuals reproduce, to propose new evolutionary processes that aren't evolutionary processes at all (but proximate behavioral patterns that *require* evolutionary explanations), and to presume without justification that certain quirks of modern social life were selection pressures of our deep evolutionary past. Explaining the exceptional features of human cooperation is challenging enough without muddling the problem space even further with conceptual false starts, questionable historical premises, and labyrinthine evolutionary scenarios.

Human evolutionary exceptionalism is counterproductive for science. It leads to internecine squabbles. Correcting the misconceptions that follow in its wake distracts specialists from more productive work. Finally, it confuses nonspecialists, who lack the time to sort through these controversies for themselves. It's good to be curious—and, sometimes, even querulous—about how our biologically exceptional traits evolved, but we should resist the idea that evolution made up new rules just for us.

ANIMAL MINDLESSNESS

KATE JEFFERY

Professor of behavioral neuroscience; head, Research Department of Cognitive, Perceptual, and Brain Sciences, University College London

We humans have had a tough time coping with our unremarkable place in the grand scheme of things. First, Copernicus trashed our belief that we live at the center of the universe. He was followed shortly thereafter by Herschel & Co., who suggested that our sun was not at the center of it either. Then Darwin came along and showed that according to our biological heritage, we're just another animal. But we have clung for dear life to one remaining belief about our specialness—that we, and we alone, have conscious minds. It's time to retire—indeed, euthanize and cremate—this anthropocentric pomposity.

Descartes thought of animals as mindless automata, and vivisection without anesthetic was common among early medical researchers. Throughout much of the 20th century, psychologists believed that animals—while clearly resembling humans in their neuroanatomy—performed their activities essentially unthinkingly, a viewpoint that reached its zenith (or my preferred word, nadir) in behaviorism, the psychological doctrine that rejects inner mental states like plans and purposes as unable to be studied, or, in the radical version, as not even existing. The undeniable fact that humans have inner mental states and purposes was attributed to our special psychological status: We have language, and therefore we're different. Animals, though, remain essentially Cartesian automata.

Many of our scientific experiments have validated this view. Rats in a Skinner box (named after the most radical behaviorist of all, B. F. Skinner) do indeed appear to act mindlessly—they press the levers over and over again, they seem slow to learn, slow to adapt to new contingencies, they don't really seem to think about what they're doing. In further testament to its mindlessness, large regions of the rat brain can be damaged without affecting performance. Rats in a maze seem similarly clueless: They take a long time to learn (weeks to months, sometimes) and a long time to adapt to change. Clearly, rats and other animals are stupid—and more than that, they're mindless.

Fond though I am of rats, I would not wish to defend their intelligence. But the assumption that they don't have inner mental states needs examining. Behaviorism arose from the argument of parsimony (Occam's razor): Why postulate mental states in animals when their behavior can be explained in simpler ways? The success of behaviorism arose in part because the kinds of behaviors studied back then could indeed be explained by operation of mindless, automatic processes. It doesn't take deep reflection to press a lever in a Skinner box, any more than it does to key in your PIN. But in the mid-20th century, a development occurred that began to overturn the view that *all* behavior is mindless. This development was single-neuron recording—the ability to follow the activity of individual brain cells, the little cogs and sprockets making up the workings of the brain. Using this technique, behavioral electrophysiologists have seen for themselves the operation of inner mental processes in animals.

The most striking of these discoveries has been the place cells—neurons in the hippocampus, a small but vitally important structure located deep in the temporal lobes. Place cells are (we now know) key components of an internal representation of the environment—often called the cognitive map—which forms when an animal explores a new place and reactivates when the animal reenters that place. Single-neuron recording shows us that this map forms spontaneously, in the absence of reward, and independently of the animal's behavior. When an animal is choosing between alternative routes to a goal, place cells representing the alternative possibilities become spontaneously active even though the animal hasn't gone there yet—as if the animal were *thinking about* the choices. Place cells certainly seem to be an internal representation. Furthermore, we humans have them too, and human place cells reactivate when people think about places.

Place cells may well be an internal representation of the kind eschewed by behaviorists, but does this mean that rats and other animals have minds? Not necessarily; place cells could still be part of an automatic and unconscious representation system. Our own ability to conjure up remembered or imagined images "in the mind's eye" to use for recollection or planning might still be special. This seems unlikely though, doesn't it? Mindlessness would be only a parsimonious conjecture if we didn't know about our own minds. But we do, and we know we're extraordinarily like animals in every respect, right down to the place cells. To suppose that the ability to mentally represent the outside world sprang into existence, fully formed, in the evolutionary transition (if the concept of "transition" even makes sense) between animals and humans seems improbable at best, deeply arrogant at worst. When we look into the animal brain, we see the same things we see in our own brains. Of course we do, because we're just animals after all. It's time to admit yet again that we're not all that special. If we have minds, creatures with brains very like ours probably do, too. Unraveling the mechanisms of these minds will be the great challenge for the coming decades.

HUMANIQUENESS

IRENE PEPPERBERG

Research associate in psychology, Harvard University; adjunct associate professor in psychology, Brandeis University; author, Alex & Me

Yes, humans do some things that other species don't—we're indeed the only species to send probes into space to find other forms of life—but the converse is equally true: Other species do things that humans find impossible, and many nonhuman species are unique in their abilities. No human can detect temperature changes of a few hundredths of a degree, as can some pit vipers, nor can humans best a dog at following faint scents. Dolphins hear at ranges impossible for humans and, along with bats, can use natural sonar. Bees and many birds see in the ultraviolet, and many birds migrate thousands of miles yearly under their own power, with what seems to be some kind of internal GPS. Humans, of course, can and will invent machines to accomplish such feats of nature, unlike our nonhuman brethren—but nonhumans had these abilities first. I don't contest data showing that humans are unique in many ways, and I certainly favor studying the similarities and differences across species, but I think it's time to retire the notion that human uniqueness is a pinnacle of some sort, denied in any shape, way, or form to other creatures.

Another reason for retiring the idea of humaniqueness as *the* ideal endpoint of some evolutionary process is, of course, that our criteria for uniqueness inevitably need redefinition. Remember when "man, the tool user" was our definition? At least until along came species like cactus-spike-using Galapagos finches, sponge-wielding dolphins, and now even crocodiles that use sticks to lure birds to their demise. Then it was "man, the tool *maker,*" but that fell out of favor when such behavior was seen in a number of other creatures, including species as evolutionarily distant from humans as New Caledonian crows. Learning through imitation? Almost all songbirds do it to some extent vocally, and minor evidence exists for physical aspects in parrots and apes. Current research does demonstrate that apes, for example, lack certain aspects of collaborative abilities seen in humans, but one wonders whether different experimental protocols may yet provide different data.

The comparative study of behavior needs to be expanded and supported, but not merely to find more data enshrining humans as "special." Finding out what makes us different from other species is a worthy enterprise, but it can also lead us to find out what is "special" about other beings—what incredible things we may need to learn from them. So, for example, we need more studies to determine the extent to which nonhumans show empathy or exhibit various aspects of "theory of mind," to learn what's needed for their survival in their natural environment and what they can acquire when enculturated into ours. Maybe they have other means of accomplishing the social networking we take as at least a partial requisite for humanness. We need to find out what aspects of human communication skills they can acquire, but we must also uncover the complexities existing in their own communication systems.

Nota bene: Lest my point be misunderstood, my argument is a different one from that of bestowing personhood on various nonhuman species and is separate from other arguments for animal rights and even animal welfare—although I can see the possible implications of what I propose.

All told, it seems time to continue studying the complexities of behavior in all species, human and nonhuman; to concentrate on similarities as well as differences; and, in many cases, to appreciate the inspiration that our nonhuman compatriots provide, in order to develop tools and skills that enhance our own abilities—rather than simply consigning nonhumans to second-class status.

HUMAN BEING = *HOMO SAPIENS*

STEVE FULLER

Philosopher, sociologist; Auguste Comte Chair in Social Epistemology, University of Warwick, U.K.; author, The Proactionary Imperative: A Foundation in Transhumanism

It's difficult to deny that humans began as *Homo sapiens,* an evolutionary offshoot of the primates. Nevertheless, for most of what is properly called "human history" (that is, the history starting with the invention of writing), most of *Homo sapiens* have not qualified as "human"—and not simply because they were too young or too disabled. In sociology, we routinely invoke a trinity of shame—race, class, and gender—to characterize the gap that remains between the normal existence of *Homo sapiens* and the normative ideal of full humanity. Much of the history of social science can be understood as either directly or indirectly aimed at extending the attribution of humanity to as much of *Homo sapiens* as possible. It's for this reason that the welfare state is reasonably touted as social science's great contribution to politics in the modern era. But perhaps membership in *Homo sapiens* is neither sufficient nor even necessary to qualify a being as "human." What happens then?

In constructing a scientifically viable concept of the human, we could do worse than take a lesson from republican democracies, which bestow citizenship on those whom its members are willing to treat as equals in some legally prescribed sense of reciprocal rights and duties. Republican citizenship is about the mutual recognition of peers, not a state of grace bestowed by some overbearing monarch. Moreover, republican constitutions define citizenship in terms that don't make explicit reference to the inherited qualities of the citizenry. Birth in the republic doesn't constitute a privilege over those who have had to earn their citizenship. A traditional expression of this idea is that those born to citizens are obliged to perform national service to validate their citizenship. The United States has exceeded the wildest hopes of republican theorists (who tended to think in terms of city-states), given its historically open-door immigration policy yet consistently strong sense of self-identity—not least among recent immigrants.

In terms of a scientifically upgraded version of "human rights" that might be called "human citizenship," let's imagine this open-door immigration policy as ontological rather than geographical in nature. Thus, non–*Homo sapiens* may be allowed to migrate to the space of the "human." Animal-rights activists believe they are already primed for this prospect. They can demonstrate that primates and aquatic mammals are not only sentient but also engaged in various higher cognitive functions, including what's nowadays called "mental time-travel." This is the ability to set long-term goals and pursue them to completion because the envisaged value of the goal overrides that of the diversions encountered along the way. While this is indeed a good empirical marker of the sort of autonomy historically required for republican citizenship, in practice animal-rights activists embed this point in an argument for *de-facto* species segregationism, a "separate but equal" policy, in which the only enforceable sense of "rights" is one of immunity from bodily harm from humans. It is the sense of "rights" *qua* dependency that a child or a disabled person might enjoy.

The fact that claims to animal rights carry no sense of reciprocal obligations on the part of the animals toward humans raises question about the activists' sincerity in appealing to "rights" at all. However, if the activists are sincere, then they should also call for a proactive policy of what the science-fiction writer David Brin has termed "uplift," whereby we prioritize research designed to enable cognitively privileged creatures, regardless of material origin, to achieve capacities enabling them to function as peers in what might be regarded as an expanded circle of humanity. Such research may focus on gene therapy or prosthetic enhancement, but in the end it would inform a Welfare State 2.0 that takes seriously our obligation to all those we regard as able to be rendered human, in the sense of fully autonomous citizens in the Republic of Humanity.

The idea that "human being" = *Homo sapiens* has always had a stronger basis in theology than biology. Only the Abrahamic religions have clearly privileged the naked ape over all other creatures. Evolutionists of all stripes have seen only differences in degree as separating the powers of living things, with relatively few evolutionists expecting that a specific bit of genetic material will someday reveal the "uniquely human." All the more reason to think that, in a future where some version of evolution prevails, republican theories of civil rights are likely to point the way forward. This prospect implies that every candidate being will need to earn the status of "human" by passing certain criteria as determined by those in the society in which he, she, or it proposes to live. The Turing Test provides a good prototype for examining eligibility for this expanded circle of humanity, given its neutrality to material substratum.

It's not too early to construct Turing Test 2.0 tests of human citizenship that attempt to capture the full complexity of the sorts of beings we would have live among us as equals. A good place to start would be with a sympathetic rendering of longstanding—and too easily dismissed—"anthropomorphic" attributions to animals and machines. Welfare State 2.0 policies could then be designed to enable a wide assortment of candidate beings—from carbon to silicon—to meet the requisite standard of citizenship implied in such attributions. Indeed, many classic welfare-state policies, such as compulsory mass education and childhood vaccination, can be understood retrospectively as the original political commitment to "uplift" in Brin's sense—but applied only to members of *Homo sapiens* living within the territory governed by a nation-state.

However, by removing the need to be *Homo sapiens* to qualify for human citizenship, we're faced with a political situation comparable to the European Union's policy for accession of new member states. The policy assumes that candidate states start with certain historical disadvantages vis-à-vis membership in the Union but that these are in principle surmountable. Thus, there's a pre-accession period in which the candidate states are monitored for political and economic stability, as well as treatment of their own citizens, after which integration occurs in stages—starting with the free mobility of students and workers, the harmonization of laws, and revenue transfers from more established member states. To be sure, there's push-back by both the established and the candidate member states. But notwithstanding these painful periods of mutual adjustment, the process has so far worked and may prove a model for the ontological union of humanity.

ANTHROPOCENTRICITY

SATYAJIT DAS

Expert, derivatives & risk; author, Extreme Money

Parallax describes the apparent change in direction of a moving object caused by alteration in the observer's position. In the graphic work of M. C. Escher, human faculties are similarly deceived and an impossible reality made plausible.

While not strictly a scientific theorem, anthropocentrism, the assessment of reality through an exclusively human perspective, is deeply embedded in science and culture. Improving knowledge requires abandoning anthropocentricity, or at least acknowledging its existence.

Anthropocentrism's limits derive from the physical constraints of human cognition and specific psychological attitudes. Being human entails specific faculties, intrinsic attitudes, and values and belief systems that shape inquiry and understanding.

The human mind has evolved a specific physical structure and biochemistry that shapes our thought processes. The human cognitive system determines our reasoning and therefore our knowledge. Language, logic, mathematics, abstract thought, cultural beliefs, history, and memories create a specific human frame of reference, which may restrict what we can know or understand.

There may be other forms of life and intelligence. The ocean has revealed creatures that live from chemo-synthesis in ecosystems around deep-sea hydrothermal vents, without access to sunlight. Life-forms based on materials other than carbon may also be feasible. An entirely radical set of cognitive frameworks and alternative knowledge cannot be discounted.

Like a train that can run only on tracks, which determine its direction and destination, human knowledge may ultimately be constrained by what evolution has made us.

Knowledge was originally driven by the need to master the natural environment to meet basic biological needs—survival and genetic propagation—and to deal with the unknown and forces beyond human control. Superstition, religion, science, and other belief systems evolved to meet these human needs. In the 18th century, medieval systems of aristocratic and religious authority were supplanted by a new model of scientific method, rational discourse, personal liberty, and individual responsibility. But this didn't change the basic underlying drivers. Knowledge is also influenced by human factors—fear, greed, ambition, submission, tribal collusion, altruism, and jealousy, as well as complex power relationships and interpersonal group dynamics. Behavioral science illustrates the inherent biases in human thought.

Understanding operates within these biological and attitudinal constraints. As Friedrich Nietzsche wrote, "Every philosophy also hides a philosophy; every opinion is also a hiding place, every word is also a mask." Understanding of fundamental issues remains limited. The cosmological nature and origins of the universe are contested. The physical source and nature of matter and energy are debated. The origins and evolution of biological life remain unresolved. Resistance to new ideas frequently restricts the development of knowledge. The history of science is a succession of controversies—a non-geocentric universe, continental drift, the theory of evolution, quantum mechanics, climate change.

Science, paradoxically, seems also to have inbuilt limits. Like an endless set of Russian dolls, quantum physics is an endless succession of seemingly infinitely divisible particles. Werner Heisenberg's uncertainty principle posits that human knowledge about the world is always incomplete, uncertain, and highly contingent. Kurt Gödel's incompleteness theorems of mathematical logic establish inherent limitations of all but the most trivial axiomatic systems of arithmetic. Experimental methodology and testing is flawed. Model predictions are often unsatisfactory. As Nassim Nicholas Taleb has observed, "You can disguise charlatanism under the weight of equations, and nobody can catch you, since there is no such thing as a controlled experiment."

Challenging anthropocentrism doesn't mean abandoning science or rational thought. It doesn't mean reversion to primitive religious dogma, messianic phantasms, or obscure mysticism. Transcending anthropocentricity may allow new frames of reference, expanding the boundary of human knowledge. It may allow human beings to think more clearly and consider different perspectives; it may encourage possibilities outside the normal range of experience and thought. It may also allow a greater understanding of our existential place within nature and in the order of things.

As William Shakespeare's Hamlet cautioned a friend, "There are more things in heaven and earth, Horatio, than are dreamt of in your philosophy." But fundamental biology may not allow the required change of reference framework. While periodically humbled by the universe, human beings remain enamored, for the most part, of the notion that they are the apogee of development. But as Mark Twain observed in *Letters from the Earth,* "He took a pride in man; man was his finest invention; man was his pet, after the housefly."

Writing in *The Hitchhiker's Guide to the Galaxy,* the late English author Douglas Adams speculated that Earth was a powerful computer and human beings were its biological components, designed by hyperintelligent pandimensional beings to answer the ultimate questions about the universe and life. To date, science has not produced a conclusive refutation of this whimsical proposition.

Whether or not we can go beyond anthropocentrism, it's a reminder of our limits. As Martin Rees, emeritus professor of cosmology and astrophysics at Cambridge, has noted:

> Most educated people are aware that we are the outcome of nearly 4 billion years of Darwinian selection, but many tend to think that humans are somehow the culmination. Our sun, however, is less than halfway through its life span. It will not be humans who watch the sun's demise, 6 billion years from now. Any creatures that then exist will be as different from us as we are from bacteria or amoebae.[ad]

TRUER PERCEPTIONS ARE FITTER PERCEPTIONS

DONALD D. HOFFMAN

Cognitive scientist, University of California, Irvine; author, Visual Intelligence

Those of our predecessors who perceived the world more accurately enjoyed a competitive advantage over their less fortunate peers. They were thus more likely to raise children and become our ancestors. We're the offspring of those who perceived more truly, and we can be confident that our perceptions are, in the normal case, reasonably accurate. There are of course endogenous limits. We can, for instance, perceive light only in a narrow window of wavelengths between roughly 400 and 700 nanometers and sound only in a narrow window of frequencies between 20 and 20,000 hertz. Moreover, we're prone on occasion to have perceptual illusions. But with these provisos noted, it's fair to conclude on evolutionary grounds that our perceptions are in general reliable guides to reality.

This is the consensus of researchers studying perception via brain imaging, computational modeling, and psychophysical experiments. It's mentioned in passing in many professional publications and stated as fact in standard textbooks. But it gets evolution wrong. Fitness and truth are distinct concepts in evolutionary theory. To specify a fitness function, you must specify not just the state of the world but also, *inter alia*, a particular organism, a particular state of that organism, and a particular action. Dark chocolates can kill cats but are a fitting gift from a suitor on Valentine's Day.

Monte Carlo simulations using evolutionary game theory, with a wide range of fitness functions and of randomly created environments, find that truer perceptions are routinely driven to extinction by perceptions tuned to the relevant fitness functions. The extension of these simulations to evolutionary graphs is in progress, and the same result is expected. Simulations with genetic algorithms find that truth never gets onstage to have a chance to go extinct.

Perceptions tuned to fitness are typically far less complex than those tuned to truth. They require less time and resources to compute and are thus advantageous in environments where swift action is critical. But even apart from considerations of time and complexity, true perceptions go extinct simply because natural selection selects for fitness, not truth. We must take our perceptions seriously. They've been shaped by natural selection to guide adaptive behaviors and keep us alive long enough to reproduce. We should avoid cliffs and snakes. But we mustn't take our perceptions *literally*. They aren't the truth; they're simply a species-specific guide to behavior.

Observation is the empirical foundation of science. The predicates of this foundation—including space, time, physical objects, and causality—are a species-specific adaptation, not an insight. Thus this view of perception has implications for fields beyond perceptual science, including physics, neuroscience, and the philosophy of science. The old assumption that fitter perceptions are truer perceptions is deeply woven into our conception of science. The funeral of this assumption won't be snubbed with a back-page obituary but heralded with regime change.

THE INTRINSIC BEAUTY AND ELEGANCE OF MATHEMATICS ALLOWS IT TO DESCRIBE NATURE

GREGORY BENFORD

Emeritus professor of physics and astronomy, University of California, Irvine; science fiction author, Shipstar

Many believe this seeming axiom—that beauty leads to descriptive power. Our experience seems to show this, mostly from the successes of physics. There's some truth to it, but also some illusion.

There's a ready explanation of how a long-ago primate acquired the beginnings of a mathematical appreciation of nature. Hunting, that primate found it easier to fling rocks or spears at fleeing prey than to chase them down. Some of his fellows found the curve of a flung stone difficult to achieve, but he didn't. He found the parabola beautiful and simpler to achieve, because that pleasurable sensation provided evolutionary feedback. Over eons, this led to an animal that invented complex geometries, calculus, and beyond.

This is a huge leap, of course—an evolutionary overshoot. We seem to be smarter than needed simply to survive in the natural world; earlier hominids did, even spreading over most of the planet. We did go through some population bottlenecks in our past, perhaps as recently as 130,000 years ago. Those eras of intense selection may explain why we have such vastly disproportionate mental abilities. Still, there remain, beyond evolutionary arguments, two mysteries in math: Whence its amazing ability to describe nature? And why its intrinsic beauty and elegance?

Parabolas are elegant, true; they describe how hard bodies fly through the air under gravity. But the motion of a falling leaf demands several differential equations, taking into account wind velocity, gravity, the leaf's geometry, fluid flow, and much else. A cruising airplane is even harder to describe. Neither case is elegant or simple. So the utility of math stands separate from its intrinsic beauty. Mathematics is most elegant when we simplify the system considered. With a baseball, we account for the initial acceleration and angle, the air, and gravity, and out comes a parabola as a good approximation. Not so with the leaf.

And that parabola? We see its simple beauty far too slowly for it to be of any use in real time. Our appreciation comes afterward. To actually make a parabola work for us in baseball, we learn how to throw. Such learning builds on hardwired neuronal networks in the brain, selected for over evolutionary times, since knowing how to throw a missile is adaptive. A human pitcher can more subtly affect the trajectory by throwing curves, knuckle balls, and so on. Those are more complex trajectories, probably less elegant but still well within the capability of our nervous system. But for well-learned actions, all that processing goes on at unconscious levels. In fact, too much conscious attention to the details of action can interfere. Athletes know this—it's the art of staying in the zone. Probably that zone is where the mind runs on its sense of rightness, beauty, economy of effort.

Further, elegance is hard to define, as are most aesthetic judgments. Richard Feynman once noted that it's simple to make known laws more elegant—say, by starting with Newton's force law, $F = ma$, then defining $R = F - ma$. The equation $R = 0$

is visually more elegant but contains no more information. The Lagrangian method in dynamics is elegant—just write the expression for kinetic energy minus the potential energy—but one must know a fundamental theory to do so. The elegance of the Lagrangian comes later, as a mathematical aid.

More recently, it's hard to devise an elegant cosmological theory that yields directly the small cosmological constant we observe. Some solve this problem by invoking the anthropic principle, and thus multiverses of some sort. But this ventures near a violation of another form of the elegance standard, Occam's razor. Imagining a vast sea of multiverses, with us arising in one where conditions produce intelligent beings, seems excessive to many of us. It invokes a plentitude we can never see. The scientific test of multiverse cosmology is whether it leads to predictable consequences.

Can multiverses converse with one another? That would be a way of verifying the basis of such theories. Most multiverse models say there's no possible communication between the infinitude of multiverses. Brane theory, though, comes from models where no force law operates between branes except gravitation. Perhaps someday an instrument like LIGO, the Laser Interferometer Gravitational-wave Observatory, can detect such waves from branes. But is it elegant to shift confirmation onto some far future technology? Sweeping dust under a rug seems inelegant to me.

Evolution doesn't care about beauty and elegance, just utility. Beauty does play a secondary role, though. The male who best throws the spear to bring down prey is appreciated and may have a choice of many mates. It just so happens that the effective and now beautiful act of spear throwing is describable with fairly simple math.

We make the short step to say the underlying math is also beautiful.

Math's utility implies that for a suitably simple model of the universe there should be a fairly simple mathematical Theory of Everything—something like general relativity, describable by a one-line equation. Searching for it on that intuitive basis may lead us to such a theory. I suspect that a model capturing the full complexity of the universe, though, would take up a lot more than one line.

When we say a math model is elegant and beautiful, we express the limits of our own minds. It's not a deep description of the world. In the end, simple models are much easier to comprehend than complex ones. We cannot expect that the path of elegance will always guarantee that we're on the right track.

GEOMETRY

CARLO ROVELLI

Theoretical physicist, Professeur de classe exceptionelle, Université de la Méditerranée, Marseille; author, The First Scientist: Anaximander and His Legacy

We will continue to use geometry as a useful branch of mathematics, but it's time to abandon the longstanding idea of geometry as the description of physical space. The idea that geometry is the description of physical space is ingrained in us and might seem hard to get rid of, but getting rid of it is unavoidable and just a matter of time. Might as well get rid of it soon.

Geometry developed at first as a description of the properties of parcels of agricultural land. In the hands of the ancient Greeks, it became a powerful tool for dealing with abstract triangles, lines, circles, and the like, and it was applied to describe paths of light and movements of celestial bodies with great efficacy. In the modern age, with Newton, it became the mathematics of physical space. This geometrization of physical space appeared to be further vindicated by Einstein, who described space (actually, spacetime) in terms of the curved geometry of Riemann. But in fact this was the beginning of the end. Einstein discovered that the Newtonian space described by geometry is in fact a field, like the electromagnetic field, and fields are nicely continuous and smooth only if measured at large scales. In reality, they're quantum entities that are discrete and fluctuating. Therefore, the physical space in which we're immersed is in reality a quantum-dynamical entity that has very little in common with what we call "geometry." It's a pullulating process of finite interacting quanta. We can still use expressions like "quantum geometry" to describe it, but the reality is that a quantum geometry is not much of a geometry anymore.

Better to rid ourselves of the idea that our spatial intuition is always reliable. The world is far more complicated (and beautiful) than a "geometrical space" and things moving in it.

CALCULUS

ANDREW LIH

Associate professor, School of Communication,
American University; author, The Wikipedia Revolution

I do not propose that we should do away with the study of change or the area under the curve, or bury Isaac Newton and Gottfried Leibniz. However, for decades now, learning calculus has been the passing requirement for entry into modern fields of study by combining the rigorous requirements of science, technology, engineering, and math. Universities still require undergraduates to take anywhere from one to three semesters of calculus as a pure math discipline, typically featuring complex math concepts uncontextualized and removed from practical applications, and heavily emphasizing proofs and theorems.

Calculus has thus become a hazing ritual for those interested in going into one of the most essential fields today: computer science. Calculus has very little relevance to the day-to-day work of coders, hackers, and entrepreneurs, yet poses a significant barrier to recruiting sorely needed candidates for today's digital workforce.

This problem is particularly urgent in the area of programming and coding. Undergraduate computer-science programs are starting to bounce back from a dearth of enrollment that plagued them in the early Internet era, but we could do a lot more to fill the ranks if we rid ourselves of the lingering view that computer science is an extension of mathematics—a view that dates from an era when computers were primarily crafted as the ultimate calculators.

Calculus remains in many curricula more as a rite of passage than for any particular need. It's one way of problem solving, and it contributes to our ability to absorb more complex concepts, but retaining it as an obstacle course that one must navigate in order to program and code is counterproductive. Leaving in this obtuse math requirement is lazy curricular thinking. It sticks with a model that weeds out people for no reason related to their ability to program.

This leads us to the question, What makes for good programmers? The answer is an ability to deconstruct complex problems into a series of smaller, doable ones, to think procedurally about systems and structures, to manipulate bits and do amazing things with them. If calculus isn't a good fit for these pursuits, what should replace it? Discrete math, combinatorics, computability, and graph theory are far more important than calculus. These are all standard and immensely necessary fields in most modern computer-science programs, but they typically come after the calculus requirement gauntlet.

People are finding other formal and peer-learning methods to pick up coding outside the higher-education environment: meetups, code-a-thons, online courses, video tutorials. Dropping the calculus requirement would bring these folks into the fold earlier and more methodically. This won't mean we'll be turning universities into trade schools. We still want our research scientists-in-training and our PhD candidates in STEM to know and master calculus, linear algebra, and differential equations. But for too long, calculus has served as a choke point for training digital-savvy self-starting innovators.

Clemson University experimented with moving calculus farther down the curriculum, not as a prerequisite but as a class in sync with the need for it in other STEM classes. A 2004 longitudinal study showed "a statistically significant improvement in retention in engineering" when it reconfigured its approach to introduce math in later semesters.[ae] We need more of these experiments and further radical curricular thinking to get past the prerequisite model that has dominated the field for decades.

How can so many people be interested in coding and programming yet not be served by our top institutions of higher learning? By treating computer science largely as a STEM discipline instead of as a whole new capability cutting across several fields, we haven't evolved with the times. The sooner we move beyond STEM-oriented thinking, the better.

COMPUTER SCIENCE

NEIL GERSHENFELD
Physicist, director, Center for Bits and Atoms, MIT; author,
Fab: The Coming Revolution on Your Desktop

Computer science is a curious sort of science—one that implicitly ignores, and even explicitly opposes, the principles of the rest of science.

There are many models of computation: imperative versus declarative versus functional languages, SISD versus SIMD versus MIMD architectures, scalar versus vector versus multicore processors, RISC versus CISC versus VLIW instruction sets. But there is only one underlying physical reality: A patch of space can contain states, which can interact and take time to transit. Anything else is a fiction.

Heroic efforts are now going into maintaining that fiction. Programming today is a bit like inhabiting the pleasure gardens in Fritz Lang's *Metropolis*, confident that the workers in the machine rooms down below will follow your instructions. Interconnect bottlenecks, cache misses, thread concurrency, data-center power budgets, and the inefficiency of parallel processors (and programmers) are rumblings of discontent from the underground.

Software doesn't have physical units like time and space, but the hardware that executes it does. The code for an application program, the executable code that it's compiled to, and the circuits that run it don't look at all like each other. When a map is zoomed, there's also a hierarchical structure from city to state to country, but the geometry of the representation isn't changed. Why do we do that for software?

I blame two people for this state of affairs, Alan Turing and John von Neumann. They're famous for what were essentially historically important hacks. Turing was interested in the question of what was computable. His namesake machine was meant to be a theoretical model, not an experimental prescription. It had a head that read and wrote symbols stored on a tape. While that might sound straightforward, it's an unphysical distinction; persistence and interaction are both properties of a physical state. This segregation of function was elaborated in the organs of von Neumann's architecture. Even though that underpins almost every computer made today, it wasn't intended to be a universal truth. Rather, it was articulated in an influential report that von Neumann wrote on programming within the limited confines of an early computer, the EDVAC.

Turing and von Neumann understood the limits of their models; late in life, they both studied computing in spatial structures—pattern formation for Turing, self-replication for von Neumann. But their legacy lives on in the instruction pointer in almost any processor, the modern descendant of Turing's head reading a tape. All the other instructions not pointed to consume information-processing resources but don't process information.

In nature, everything happens everywhere all the time. While an industry has developed devices *for* computation, a much smaller community has studied the physics *of* computation. Outside of what is traditionally considered to be computer science, they have developed quantum computers that use entanglement and superposition, microfluidic logic that transports material as well as information, analog logic that solves digital problems with continuous device degrees of freedom, and digital fabrication to code construction of programmable materials. Most important, programming models are emerging that represent and respect physical resources, rather than viewing them as a can to be kicked to someone else to worry about. This is turning out to be easier rather than harder to do, because it avoids all the issues of converting from an unphysical to a physical world.

In the movie *The Matrix*, Neo is given a choice between a red pill to exit the fictional world he's been inhabiting or a blue pill to maintain the illusion. What he found when he got out was much messier but ultimately much more satisfying. There's a similar choice now before the digital world—between avoiding or embracing the physical reality it inhabits.

Think of Turing's machine and von Neumann's architecture as technological training wheels. They've given us a good ride, but something of a do-over is now needed: to introduce physical units into software in order to be able to program the ultimate universal computer, the universe.

SCIENCE ADVANCES BY FUNERALS

SAMUEL BARONDES

Jeanne and Sanford Robertson Chair in Neurobiology and Psychiatry, University of California, San Francisco; author, Making Sense of People

When Max Planck began studying physics at the University of Munich in 1874, his teacher, Philipp von Jolly, warned him that it was already a mature field, with little more to learn. This attitude was widely held through the end of the 19th century. In 1900, Lord Kelvin, the great British physicist, put it clearly: "There is nothing new to be discovered in physics now. All that remains is more and more precise measurement."

In Planck's early career, he had no reason to doubt this complacent position. And yet in the same year that Kelvin made his pronouncement, Planck found himself disproving it. He had been working on the relationship of heat to light, a topic of great interest to the emerging electric companies, and he had proposed an equation consistent with classical physical concepts. But he was dismayed to learn of new experimental results that proved him wrong.

With his back against the wall, the forty-two-year-old Planck quickly thought up an alternative equation that fit the data. But the new equation also had a disruptive effect. Hard to reconcile with traditional ideas, it turned out to be an initial building block for a completely new view of physics called quantum theory. The resistance to this disruption by conservative members of the physics community may be taken as one example of Planck's petulant claim that a new scientific truth will not triumph until "its opponents eventually die."

But the triumph of quantum theory didn't really depend on this grim prospect. Members of the physics establishment soon began to take quantum theory seriously, because it wasn't just a weird idea that had popped into Planck's head. It had become necessary because of a surprising experimental result. This is how science usually works. When experiments challenge a prevailing idea, attention is paid. If the experiments are confirmed, the old idea is modified. In fields in which decisive experimentation is relatively easy, such change may happen quickly and is certainly not dependent on the death of its senior practitioners. Only in fields that don't lend themselves to decisive experimentation is it hard to definitively challenge a prevailing position. In such fields, even death may not be enough, and tenuous positions may survive for generations.

So Planck got it wrong. The development of new scientific truths doesn't depend on the passing of stubborn conservative opponents. It is, instead, mainly dependent on the continuous enrollment of talented newcomers eager to make their mark by changing the existing order. In Planck's case, it was the arrival of the young Albert Einstein, rather than the demise of his senior opponents, that propelled quantum theory forward. As Douglas Stone showed in *Einstein and the Quantum*, it was the twenty-five-year-old patent clerk, a fledgling outsider with nothing to lose, who became the driving force in the development of this theory. As for his elders, Einstein couldn't have cared less.

PLANCK'S CYNICAL VIEW OF SCIENTIFIC CHANGE

HUGO MERCIER

Cognitive scientist; Ambizione Fellow,
Cognitive Science Center, University of Neuchâtel

This year's *Edge* Question was inspired by Max Planck's bleak view of scientific change: "A new scientific truth does not triumph by convincing its opponents and making them see the light, but rather because its opponents eventually die, and a new generation grows up that is familiar with it." Certainly, Planck's assessment struck a chord with the general public. Its reception among the more educated public was likely eased when Thomas Kuhn pointed out that well-established scientists would have an incentive to resist novel theories instead of jettisoning their life's work.

If even scientists, with their freedom of discourse and exacting standards of evidence, cannot change their minds when they should, what hope is there for the rest of us? Why bother trying to convince anyone, ever?

Fortunately, Planck was wrong. Detailed accounts of major scientific changes reveal, time after time, how quickly scientists adopt novel theories—provided they're well supported.

One can hardly blame, for instance, 16th-century scholars for rejecting Copernicus' heliocentric model: It didn't account for the data much better than the alternatives, it was laden with inelegant *post-hoc* fixes, and it had no answer to such basic question as, "If the Earth is moving, then why can't we feel it?" As these issues got resolved—Kepler introduced elliptical orbits;

Galileo understood the principles of motion—the heliocentric model promptly gained supporters.

Other theories that also required dramatic conceptual change were much more quickly accepted, as they rested from the start on better arguments. When Newton first advanced a new theory of light, one that upset centuries-old beliefs, he did so in a short article that offered little experimental evidence for many of his claims. Yet the cogency of his theory proved persuasive to many (this was not a case of argument from authority, since Newton had very little at the time). When, thirty years later, Newton published his *Opticks*, with a much better presentation of the same theory and a plethora of well-described experiments, he took natural philosophers by storm; a few years and a few replications later, most were sold on his ideas.

By taking his belief in the existence of phlogiston to the grave, Joseph Priestley became a favorite example of the pig-headedness of even brilliant scientists. But Priestley was very much an exception. When Lavoisier started publicizing his discoveries and criticizing the concept of phlogiston, he was met with resistance but also with acceptance—resistance to new theories that were half-baked even in Lavoisier's own mind, acceptance for his solid methods and results. Once the French chemist formulated a theory that could properly account for the main phenomena of interest, it was accepted in a matter of years.

Examples multiply: The heart of Darwin's ideas was accepted by his colleagues shortly after publication of the *Origin*, plate tectonics went from speculation to textbook example in a dozen years—both showing that when the arguments are good, the vast majority of scientists change their minds accordingly. As science historian Bernard Cohen has noted, even Planck, whose ideas were no less revolutionary than the examples mentioned here, managed to convince most of his peers, not only the new generation.

Evidently not every science reaches a consensus equally quickly—political scientists, say, do not have the benefit of data quite as precise as that gathered by particle physicists. Still, it's important to give science as a whole its due, not only because such efficient belief change is no mean feat but also because a pessimistic, cynical view of the power of argumentation can have pernicious effects.

If people who disagree with us are never going to change their minds, then why even talk to them? If we don't engage in discussion with people who disagree with us, we'll never learn of the (often perfectly good) reasons they disagree with us. If we cannot address those reasons, our arguments are likely to prove unconvincing. Our failures to convince will only reinforce the belief that we face pigheadedness rather than rational disagreement. A belief in the inefficiency of argumentation can be a destructive, self-fulfilling prophecy. We should give scientists, and argumentation generally, more credit. Let's retire Planck's cynical view of scientific change.

NEW IDEAS TRIUMPH BY REPLACING OLD ONES

JARED DIAMOND

Professor of geography, University of California, Los Angeles; author, The World Until Yesterday

The history of science is much more variegated than assumed in the *Edge* Question about the abandonment and burial of old ideas. While the view that new ideas triumph by replacing old ones fits some scientific developments, in many other cases new ideas take over a vacuum formerly occupied by no well-articulated idea at all. That happens for either of two reasons: new ideas responding to new information made possible by new measurements, or else responding to new "outlooks." (Among historians of science, the term used rather than the inadequate English term "outlook" is the German *Fragestellung*—literally, the posing of a question, but more broadly meaning a world-view from which that question can arise.) I'll give two or three examples illustrating each of those two reasons.

The most familiar modern example of a new idea made possible by new measurements is Watson and Crick's double-helix model of the structure of DNA. Their model didn't replace a previous established model whose opponents gradually died out without abandoning their error. Instead, the Watson-Crick model was made possible by two recent sets of measurements: analyses of the chemical composition of DNA (revealing equivalent amounts of the bases adenine and thymine and of cytosine and guanine); and X-ray crystallographic evidence. As is well known, two models of DNA structure were then proposed nearly simultaneously, by Pauling and by Watson and Crick. It almost immediately became obvious that the former model was wrong and the latter model accounted for all the evidence. Hence the Watson-Crick model became rapidly accepted, replacing a vacuum rather than a previous wrong theory.

My other example of an idea made possible by new measurements concerns the origins of animal electricity. Our nerve and muscle membranes operate by conducting electrical impulses, arising from a change in transmembrane voltage between active and inactive membrane regions. In the absence of direct measurements of transmembrane voltage, it was impossible to propose a quantitative theory for how that voltage could change. That problem was solved between 1939 and 1952 by two developments: the anatomist J. Z. Young discovered giant nerves in squid, and physiologists developed microelectrodes small enough to insert into squid giant nerves without damaging them. Between 1945 and 1952, the physiologists Alan Hodgkin and Andrew Huxley took advantage of that anatomical discovery and that technical development to measure the electric currents moving across squid nerve as a combined function of voltage and time, and thereby to reconstruct quantitatively and in detail how a nerve impulse arises from changes in nerve membrane permeability to the positively charged ions sodium and potassium. The Hodgkin-Huxley theory was rapidly accepted because it was so convincingly correct and because it had no serious competitors. When I was a physiology student in the 1950s and 1960s, the only resistance to the theory that I recall involved some concern by nonphysiologists about whether microelectrodes were causing damage to nerve membranes (a concern answered by several types of control experiments), and a nonquantitative proposal that nerve membranes and synapse or junction membranes undergo the same permeability changes (it turned out they don't).

As for new ideas made possible by a new *Fragestellung,* consider first the foundation of the several modern sciences constituting population biology: taxonomy/systematics, evolutionary biology, biogeography, ecology, animal behavior, and genetics. At least until recently, most research in all those fields except genetics involved observations, counting, and measurements requiring no equipment. Most of that research could have been done by Aristotle, Herodotus, and their contemporaries in classical-age Greece over 2,000 years earlier. The Greeks were eminently capable of patient, accurate, quantitative observations of planets and other features of the natural world. Aristotle could similarly have examined Greek animals and plants and arrived at Linnaeus' hierarchical taxonomy. Herodotus could have compared the species of the Black Sea with those of Egypt and thereby founded biogeography. And any ancient Greek could have grown and counted pea varieties as did Gregor Mendel in the 1860s, noticed the differences between Willow Warblers and Chiffchaffs (a related warbler species) as did Gilbert White in the 1780s, watched young geese as did Konrad Lorenz in the 1930s, and thereby founded genetics, ecology, and animal behavior. But ancient Greeks lacked the necessary *Fragestellung* that lent interest to counting pea varieties and scrutinizing warblers and young geese. The rise of those branches of population biology from the 1700s onward was due to a modern *Fragestellung* that generated data (without the need for invention of microelectrodes or X-ray crystallography), which in turn generated ideas, in areas where previously there had been neither data nor detailed ideas.

Without going into specifics, I'll mention two other examples of important broad fields that arose only in recent centuries without any need for specialized technology, and that the ancients could have developed but didn't because they lacked a relevant *Fragestellung*. The Greeks and Romans were in contact with speakers of Indo-European and Semitic and other languages, could have discovered the groupings of languages in those language families, and could thereby have developed the ideas of historical linguistics—but they didn't even bother to record words of their Egyptian, Gaulish, and other subjects. In all of classical Greek and Roman literature I'm unaware of a single wordlist recorded for any "barbarian" language, in contrast to the wordlists that European travelers began routinely to gather among non-European peoples from the 1600s onward. The Greeks and Romans could equally well have noticed the observational evidence used by Freud to explore the unconscious within us—but they didn't.

All of this is not to say that the view underlying this year's *Edge* Question is always wrong. Examples in the fields in which I work myself include: the replacement of biogeographic theories assuming a static Earth by the acceptance of continental drift, from the 1960s onward; the rise of the taxonomic approach termed cladistics at the expense of previous taxonomic approaches, also from the 1960s onward; and the rise in the 1960s and 1970s, followed by the virtual disappearance, of attempts to make use of irreversible (non-equilibrium) thermodynamics in the fields of population biology and cell physiology. Instead, my main point is that the development of science follows much more diverse courses than only or predominantly the course of abandoning old ideas.

MAX PLANCK'S FAITH

MIHALY CSIKSZENTMIHALYI

Psychologist; codirector, Quality of Life Research Center, Claremont Graduate University; author, Flow

Note that in the introduction to this year's *Edge* Question, Max Planck speaks of scientific truths "triumphing." Truths don't triumph, the people who propose them do. What needs to be retired is the faith that what scientists say are objective truths, with a reality independent of scientific claims. Some are indeed true, but others depend on so many initial conditions that they straddle the boundary between reality and fiction.

A good chess move allows a player to triumph over his opponent. Does that mean that the move is triumphant? Maybe it is, in chess. We can only wish that the triumphs of science will be as innocuous.

THE ILLUSION OF CERTAINTY

MARY CATHERINE BATESON

Cultural anthropologist; professor emerita, George Mason University; visiting scholar, Sloan Center on Aging and Work, Boston College; author, Composing a Further Life

Scientists sometimes resist new ideas and hang on to old ones longer than they should, but the real problem is the public's failure to understand that the possibility of correction or disproof is a strength, not a weakness. We live in an era when it's increasingly important that the voting public be able to evaluate scientific claims and make analogies between different kinds of phenomena. But this can be a major source of error; the process by which scientific knowledge is refined is largely invisible to the public. The truth-value of scientific knowledge depends on its openness to correction, yet we all carry around ideas that science has long since revised—and we're disconcerted when asked to abandon them. Surprise! You won't necessarily drown if you go swimming after lunch.

A blatant example is the role of competition in evolution, which is treated by many as a scientifically established law of nature and often taken for granted by economists and psychologists—while others argue that evolution, being a "theory," is no more than a "guess." Biology has increasingly recognized the importance of symbiosis in evolution, alongside competition, as well as diversification that bypasses competition. But the "survival of the fittest," a metaphor drawn by Darwin from the description of early industrial society by Herbert Spencer, survives as a binding metaphor for human behavior.

Most people are uncomfortable with the notion that knowledge can be authoritative, can call for decision and action, and yet be subject to constant revision, because they tend to think of knowledge as additive, not recognizing the necessity of reconfiguring in response to new information. It is precisely this characteristic of scientific knowledge that encourages the denial of climate change and makes it so difficult to respond to what we do know in a context where much is still unknown.

What kind of evidence will convince the doubters of the reality of what might best be called "climate disruption"? Perhaps the exploration of scientific ideas in need of retirement should be an annual event, with an emphasis on the fact that each new synthesis of complex data is potentially more inclusive. Retiring concepts that no longer fit is not primarily a matter of eliminating error but of integrating new information and newly recognized connections into our understanding.

THE PURSUIT OF PARSIMONY

JONATHAN HAIDT

Social psychologist; professor, NYU Stern School of Business; author, The Righteous Mind

There are many things in life that are good to have yet bad to pursue too vigorously. Money, love, and sex, for example. I'd like to add parsimony to that list.

William of Ockham was a 14th-century English logician who said, "Entities must not be multiplied beyond necessity." That principle—now known as Occam's razor—has been used for centuries by scientists and philosophers as a tool to adjudicate among competing theories. Parsimony means "frugality" or "stinginess," and scientists should be "stingy" when building theories; they should use as little material as possible. If two theories do exactly as good a job of explaining the empirical evidence, then you should pick the simpler theory. If Copernicus and Ptolemy can both explain the movements of the heavens, including the occasional backward motion of some planets, then go with Copernicus' far more parsimonious model.

Occam's razor is a great tool when used as originally designed. Unfortunately, many scientists have turned this simple tool into a fetish. They pursue simple explanations of complex phenomena as though parsimony were an end in itself rather than a tool to be used in the pursuit of truth.

The worship of parsimony is understandable in the natural sciences, where it sometimes does happen that a single law or principle, or a very simple theory, explains a vast and diverse set of observations. Newton's three laws really do explain the movements of all inanimate objects. Plate tectonics really does explain earthquakes, volcanoes, and the complementary coastlines of Africa and South America. Natural selection really does explain why plants, animals, and fungi look as they do.

But in the social sciences, the overzealous pursuit of parsimony has been a disaster. Since the 18th century, some intellectuals have striven to do for the social world what Newton did for the physical world. Utilitarians, the French *philosophes*, and other utopian dreamers longed for a social order based on rational principles and a scientific understanding of human behavior. Auguste Comte, one of the founders of sociology, originally called his new discipline "social physics."

And what do we have to show for 250 years of pursuit? We have a series of time-wasting failures and ideological battles. Not all human behavior can be explained by positive and negative reinforcement (*contra* the behaviorists). Nor is it all about sex, money, class, power, self-esteem, or even self-interest—to name some of the major explanatory idols worshipped in the 20th century.

In my own field, moral psychology, we have suffered from the same overzealous pursuit of parsimony. The Harvard psychologist Lawrence Kohlberg said that morality was all about justice. Others say it's about compassion. Others say morality is all about forming coalitions, or preventing harm to others. But in fact morality is complicated, pluralistic, and culturally variable. Human beings are products of evolution, so the psychological foundations of morality are innate (as I and many others have argued at Edge.org in recent years). But there are many of these foundations, and they're just the beginning of the story. You must still explain how morality develops in such variable ways—around the world and even among siblings within a single family.

The social sciences are hard because human beings differ fundamentally from inanimate objects. People insist on making or finding meaning in things. They do it collectively, creating baroque cultural landscapes that can't be explained parsimoniously, and they do it individually, creating their own unique symbolic worlds nested within their broader cultures. As the anthropologist Clifford Geertz put it, "Man is an animal suspended in webs of significance he himself has spun." This is why it's so hard to predict what any individual will do. This is why there are almost no equations in psychology or sociology. This is why there will never be a Newton in the social sciences.

Let's retire the pursuit of parsimony from the social sciences. Parsimony is beautiful when we find it, but the pursuit of parsimony is sometimes an obstacle to the pursuit of truth.

THE CLINICIAN'S LAW OF PARSIMONY

GERALD SMALLBERG

Practicing neurologist, New York City; playwright, Off-Off Broadway Productions, Charter Members, The Gold Ring

The law of parsimony, also known as Occam's razor, doesn't warrant a funeral, but it does have some problems in its description of reality. This law states that the most simple of two competing theories should be the preferred one, and that entities should not be multiplied needlessly. It maintains a lofty stature in philosophy and science and is often utilized as a literary device. Using the law of parsimony is the essence of good detective fiction, perhaps best achieved by Arthur Conan Doyle, a physician, who sharpened to perfection Occam's razor in the reasoning employed by his renowned creation, Sherlock Holmes. One of Holmes' most noted rules is that "When you have eliminated the impossible, whatever remains, however improbable, must be the truth."

As an absolute, the law of parsimony is floundering. Not because it's aging poorly but because it's increasingly challenged by the complexity of the real world. From my vantage point as a physician in the practice of clinical neurology, its usefulness, which has always been a guiding principle for me, can easily lead to blind spots and errors in judgment when rigidly followed.

A case in point is that of a seventy-nine-year-old woman who complained of difficulty with her balance, with several recent falls. This could be attributed simply to age; however, there were other factors in her history that needed to be taken into account, including a diabetic neuropathy causing her feet to lose sensation, as well as compression of her cervical spinal cord, which produced weakness in her legs. She also had a hearing problem, with a long history of intermittent vertigo. Moreover, she was of Scandinavian descent, thus somewhat more prone genetically to vitamin–B12 deficiency secondary to poor absorption that in her case may have been exacerbated by medications for inhibiting acid reflux. The vitamin deficiency, by itself, can produce neuropathy and degeneration of the spinal cord.

It is in this complicated clinical setting that the law of parsimony utterly fails, and I doubt that even the great Holmes, who has the luxury of being a fictional character, could tie all these loose ends into a simple knot. To provide the appropriate care to this patient, I needed to utilize Hickam's dictum, the medical profession's counterargument to Occam's razor. This maxim of Dr. John Hickam, who died in 1970, states very simply, "A patient can have as many diagnoses as [she] damn well pleases."

The crucial role that the law of parsimony plays in how we reason is beyond question. This law dates back to the Greek philosophers, who refined it from their antecedents, since I suspect we evolved to seek simplicity over complexity. The desire for unity and singleness is satisfying and very seductive. However, at times it needs to be challenged by Hickam's dictum, which is a variation of the principle of plenitude. This view of reality also dates to ancient Greek philosophy, which postulates that if the universe is to be as perfect as possible, it must be as full as possible, in the sense that it contains as many kinds of things as it possibly could contain.

Given the complexity, inconsistency, ambiguity, and ultimate uncertainty that define our reality, we shouldn't limit ourselves to using only one or the other of these valuable tools of analysis. We need to be more willing to have our positions challenged, striving to keep an open mind to other arguments, other viewpoints, and conflicting data. In order to make the best decisions for the best reasons, we must choose the appropriate heuristics, coupled with intellectual honesty to guide our thinking as we grapple with the cunning machinations of the world we inhabit.

ESSENTIALIST VIEWS OF THE MIND

LISA BARRETT

University Distinguished Professor of Psychology, Northeastern University; research scientist, neuroscientist, Massachusetts General Hospital/Harvard Medical School

Essentialist thinking is the belief that familiar categories—dogs and cats, space and time, emotions and thoughts—each have an underlying essence that makes them what they are. This belief is a barrier to scientific understanding and progress. In pre-Darwinian biology, for example, scholars believed that each species had an underlying essence or physical type, and variation was considered error. Darwin challenged this essentialist view, observing that a species is a conceptual category containing a population of varied individuals, not erroneous variations on one ideal individual. Even as Darwin's ideas became accepted, essentialism held fast, as biologists declared that genes are the essence of all living things, fully accounting for Darwin's variation. Nowadays we know that gene expression is regulated by the environment, a discovery that—after much debate—prompted a paradigm shift in biology.

In physics, before Einstein, scientists thought of space and time as separate physical quantities. Einstein refuted that distinction, unifying space and time and showing that they're relative to the perceiver. Even so, essentialist thinking is still seen every time an undergraduate asks, "If the universe is expanding, what is it expanding *into*?"

In my field of psychology, essentialist thought still runs rampant. Plenty of psychologists, for example, define emotions as behaviors (for example, a rat freezes in fear or attacks in anger), each triggered automatically by its own circuit, so that the circuit for the behavior (freezing, attacking) is the circuit for the emotion (fear, anger). When other scientists showed that in fact rats have varied behaviors in fear-evoking situations—sometimes freezing but other times running away or even attacking—this inconsistency was "solved" by redefining fear as having multiple types, each with its own essence. This technique of creating ever finer categories, each with its own biological essence (rather than abandoning essentialism, as Darwin and Einstein did) is considered scientific progress. Fortunately, other approaches to emotion have arisen—approaches that don't require essences. Psychological construction, for example, considers an emotion like fear or anger as a category with diverse instances, just as Darwin did with species.

Essentialism can also be seen in studies that scan the human brain trying to locate the brain tissue dedicated to each emotion. At first, scientists assumed that each emotion could be localized to a specific brain region (for example, fear occurs in the amygdala), but they found that each region is active for a variety of emotions. Since then, scientists, rather than abandoning essentialism, have been searching for the brain essence of each emotion in dedicated brain networks and in probabilistic patterns across the brain, always on the assumption that each emotion has an essence to be found.

The fact that different brain regions and networks show increased activity during different emotions is not a problem just for emotion research. They also show increased activation during other mental activities, such as cognition and perception, and have been implicated in mental illnesses from depression to schizophrenia to autism. This lack of specificity has led to claims (in news stories, blogs, and popular books) that we've learned nothing from brain-imaging experiments. This seeming failure is actually a success. The data are screaming that essentialism is wrong: Individual brain regions, circuits, networks, and even neurons are not single-purpose. The data point to a new model of how the brain constructs the mind. Scientists understand data through the lens of their assumptions, however. Until these assumptions change, scientific progress will be limited.

Some topics in psychology have advanced beyond essentialist views. Memory, for example, was once thought to be a single process and later was split into distinct subtypes, like semantic memory and episodic memory. Memories are now thought to be constructed within the brain's functional architecture and not to reside in specific brain tissue. One hopes that other areas of psychology and neuroscience will soon follow suit. Cognition and emotion are still considered separate processes in the mind and brain, but there's growing evidence that the brain does not respect this division—that every psychological theory in which emotions and cognitions battle each other, or in which cognitions regulate emotions, is wrong.

Ridding science of essentialism is easier said than done. Consider the simplicity of this essentialist statement from the past: "Gene X causes cancer." It sounds plausible and takes little effort to understand. Compare this to a more recent explanation: "A given individual in a given situation, who interprets that situation as stressful, may experience a change in his sympathetic nervous system that encourages certain genes to be expressed, making him vulnerable to cancer." The latter explanation is more complicated but more realistic. Most natural phenomena don't have a single root cause. Sciences still steeped in essentialism need a better model of cause and effect, new experimental methods, and new statistical procedures to counter essentialist thinking.

Adherence to essentialism also has serious practical impacts on national security, the legal system, the treatment of mental illness, the toxic effects of stress on physical illness. . . . The list goes on. Essentialism leads to simplistic "single cause" thinking, whereas the world is a complex place. Research suggests that children are born essentialists (what irony!) and must learn to overcome it. It's time for scientists to overcome it as well.

THE DISTINCTION BETWEEN ANTISOCIALITY AND MENTAL ILLNESS

ABIGAIL MARSH

Associate professor of psychology, Georgetown University

The scientific studies of mental illness and antisocial behavior continue to occupy largely separate intellectual domains. Although some patterns of persistent antisocial behavior are nominally accorded diagnostic labels, such as "antisocial personality disorder" or "conduct disorder," the default approach to individuals who engage in persistent antisocial behavior is to view their patterns of behavior through a moral lens (as badness) rather than through a mental-health lens (as madness).

In some sense, this distinction represents progress. As recently as the 19th and early 20th centuries, individuals affected by all manner of psychopathologies were routinely confined and in some cases punished or even executed. Along with the emergence of understanding that symptoms of mental illness reflect disease processes, the emphasis has shifted toward prevention and treatment. However, this shift has not applied equally to all forms of psychopathology. For example, disorders primarily characterized by internalizing symptoms (persistent distress or fear, self-injuring behaviors)—as opposed to externalizing symptoms (persistent anger or hostility, antisocial and aggressive behaviors)—are strikingly similar in many respects: prevalence; parallel etiologies and risk factors; and detrimental effects on social, educational, and vocational outcomes. But whereas copious scientific resources are aimed at identifying the causes and disease processes of internalizing symptoms and developing therapies for them, the emphasis for externalizing symptoms remains primarily on confinement and punishment, with relatively few resources devoted to identifying causes and disease processes or developing therapies. Comparisons of federal mental-health funding, clinical trials, available therapeutic agents, and publications in biomedical journals directed toward internalizing versus externalizing symptoms all confirm this pattern. This asymmetry probably results from multiple causes, including cognitive and cultural biases that influence the decisions of scientists and policy makers, eroding support for the study of antisociality as a form of mental illness.

Cognitive biases include the tendency to view actions that harm others as more intentional and blameworthy than identical actions that happen not to result in harm to others (as has been shown by the experimental philosopher Joshua Knobe and others investigating the "side-effect effect") and to view agents who cause harm as more capable of intentional and goal-directed behavior than those who incur harm (as has been shown by the psychologist Kurt Gray and others investigating the distinction between moral agents and moral patients). These biases dictate that an individual predisposed, as a result of genetic and environmental risk factors, to behavior that harms others will be seen as more responsible for his or her behaviors than an individual predisposed to self-destructive behavior as a result of similar risk factors. Viewing those who harm others as responsible for their actions (and thus blameworthy) may reflect evolved tendencies to reinforce social norms by blaming and punishing wrongdoers for their misbehavior.

Related to these cognitive biases are cultural biases that hold self-interested behavior to be normative. Individualistic cultures view self-interest as humans' cardinal motive, superseding all other motives and underlying all human behavior. This norm may reflect the dominance of rational-choice theories of human behavior favored in economics and having many adherents among scholars in other academic domains, including psychology, biology, and philosophy. Belief in the norm of self-interest is also widespread among the public, rendering behavior that isn't self-interested inherently non-normative, or "abnormal." This may explain why behaviors and patterns of thinking that cause oneself harm or distress are seen as irrationality and mental illness, whereas otherwise similar behaviors and patterns of thinking that cause others harm or distress are seen as rational, if immoral, choices. Indeed, if the harm to others benefits the self, such behaviors may even be seen as hyperrational.

The United States is an unusually individualistic country, which may help to explain its unusually strong adherence to the norm of self-interest and perhaps also its unusually punitive (rather than treatment-focused) approach to crime and aggression. This approach can be contrasted with that of, for example, the relatively less individualistic Scandinavian nations, which emphasize treatment rather than punishment even of serious criminal offenders. Mental-health-focused approaches may reduce recidivism, a further indication that externalizing behaviors, including crime and aggression, may be most effectively considered symptoms of psychopathology in need of treatment rather than simple failures of impulse control in need of punishment—and that the distinction between antisociality and mental illness should be abandoned.

REPRESSION

DAVID G. MYERS

Social psychologist, Hope College; author, Psychology, *10th edition*

In today's Freud-influenced popular psychology, repression remains big. People presume, for example, that unburying repressed traumas is therapeutic. But do we routinely exile our painful memories? "Traumatic memories are often repressed," agree four in five undergraduates and members of the American and British general public in recent surveys reported by a University of California, Irvine research team.

Actually, say today's memory researchers, there's little evidence of such repression and much evidence of its opposite. Traumatic experiences—witnessing a loved one's murder, being terrorized by a hijacker or a rapist, losing everything in a natural disaster—rarely get banished into the unconscious, like a ghost in a closet. Traumas more commonly are etched on the mind as persistent, haunting memories. Moreover, extreme stress and its associated hormones enhance memory, producing unwanted flashbacks that plague survivors. "You see the babies," said one Holocaust survivor. "You see the screaming mothers. . . . It's something you don't forget."

The scientist/therapist "memory war" lingers, but it's subsiding. Today's psychological scientists appreciate the enormity of unconscious, automatic information-processing even as mainstream therapists and clinical psychologists report increasing skepticism toward repressed and recovered memories.

MENTAL ILLNESS IS NOTHING BUT BRAIN ILLNESS

JOEL GOLD

Clinical associate professor of psychiatry, NYU Langone Medical Center; coauthor (with Ian Gold), Suspicous Minds

IAN GOLD

Canada Research Chair in Philosophy and Psychiatry, McGill University; coauthor (with Joel Gold), Suspicous Minds

In 1845, Wilhelm Griesinger, author of the most important psychiatry textbook of the day, wrote: "What organ must necessarily and invariably be diseased where there is madness? . . . Physiological and pathological facts show us that this organ can only be the brain; . . ."[af] Griesinger's truism is regularly reiterated in our own time, because it expresses the basic commitment of contemporary biological psychiatry.

The logic of Griesinger's argument seems unassailable: Severe mental illness has to originate in a physiological abnormality of some part of the body, and the only plausible candidate location is the brain. Since the mind is nothing over and above the activity of the brain, the disordered mind is nothing more than a disordered brain. True enough. But that's not to say that mental disorders can, or will, be described by genetics and neurobiology. Here's an analogy. Earthquakes are nothing over and above the movements of a vast number of atoms in space, but the theory of earthquakes concerns tectonic plates and says nothing at all about atoms. The best scientific explanation of a phenomenon depends on where human beings find comprehensible patterns in the universe and not on how the universe is constituted. God may understand earthquakes and mental illness in terms of atoms, but we may not have the time or the intelligence to do so.

It's not a radical idea that understanding and treating brain disorders sometimes has to move outside the skull. Your heart hurls an embolus into your brain. You might now be unable to produce or understand speech, move one half of your body, or see half the world in front of you. You've had a stroke, and your brain is now damaged. The cause of your brain illness didn't originate there but in your heart. Your physicians will do what they can to limit further damage to your brain tissue and perhaps even restore some of the function lost due to the embolism. But they'll also try to diagnose and treat your cardiovascular disease. Are you in atrial fibrillation? Is your mitral valve prolapsed? Do you need a blood thinner? And they won't stop there. They'll want to know about your diet, exercise regimen, cholesterol level, and any family history of heart disease.

Severe mental illness is also an assault on the brain. But like the embolus, it may sometimes originate outside the brain. Indeed, psychiatric research has already given us clues suggesting that a good theory of mental illness will need concepts that make reference to things outside the skull. Psychosis provides a good example. A family of disorders, psychosis is marked by hallucinations and delusions. The central form of psychosis, schizophrenia, is *the* psychiatric brain disease *par excellence*. But schizophrenia interacts with the outside world—in particular, the social world. Decades of research has given us robust evidence that the risk of developing schizophrenia goes up with experience of childhood adversity, like abuse and bullying. Immigrants are at about twice the risk, as are their children.

And the risk of illness increases in a near-linear fashion with the population of your city and varies with the social features of neighborhoods. Stable, socially coherent neighborhoods have a lower incidence than neighborhoods that are more transient and less cohesive. We don't yet understand what it is about these social phenomena that interacts with schizophrenia, but there's good reason to think they're genuinely social factors.

Unfortunately, the environmental determinants of psychosis go largely ignored, but they provide opportunities for useful intervention. We don't yet have a genetic therapy for schizophrenia, and antipsychotic drugs can be used only after the fact and aren't nearly as good as we'd like them to be. Although the Decade of the Brain produced a great deal of important research into brain function and the NIH's BRAIN (Brain Research through Advancing Innovative Neurotechnologies) Initiative will do so too, almost none of our efforts have yet helped (or are likely to help) patients suffering from mental illness or those who treat them. However, reducing child abuse and improving the quality of the urban environment might well prevent some people from ever developing a psychotic illness at all.

Whatever it is about the social determinants of psychosis that make them risk factors, they must have some downstream effect on the brain (otherwise they wouldn't increase the risk of schizophrenia), but they themselves are not neural phenomena any more than smoking is a biological phenomenon because it's a cause of lung cancer. The theory of schizophrenia will have to be more expansive, therefore, than the theory of the brain and its disorders.

That a theory of mental illness should involve the world outside the brain is no more surprising than that the theory of cancer has to involve cigarette smoke. Yet what is commonplace in cancer research is radical in psychiatry. The time has come to expand the biological model of psychiatric disorder to include the context in which the brain functions. In understanding, preventing, and treating mental illness, we will rightly continue to look into the neurons and DNA of the afflicted and unafflicted. To ignore the world around them would be not only bad medicine but bad science.

PSYCHOGENIC ILLNESS

BEATRICE GOLOMB

Professor of medicine, University of California, San Diego

For entrée to the mindset behind "psychogenic illness," one need go no further than the humble hiccup. Consider the published case of the thirty-one-year-old epileptic retarded institutionalized male with refractory hiccup so severe that his hiccups "caused" melena—doctor lingo for "black tarry stool," from bleeding high in the gastrointestinal (GI) tract. (The blood oxidizes in transit, turning dark.) When a tube was inserted in his nose for some incidental reason, the man's hiccups ceased. Clearly, punishment had cured the hiccups, from which it followed that their cause was psychogenic. Thereafter, each time the poor man's hiccups commenced, attendants first menaced him with a nasogastric tube (which never worked), then molested the back of his throat with it, reliably aborting the hiccup bouts. After some months of nasopharynx torment, the man's hiccups resolved, and so did the melena, proving their punishment worked.

Except hiccups don't cause melena, and nasopharyngeal stimulation doesn't cure hiccups by punishment. Manifestly, a GI woe caused the melena, simultaneously irritating the input arm to the hiccup reflex—the vagus nerve, which traverses the GI tract. GI afflictions are the chief cause of persistent hiccup, and nasopharynx stimulation the most effective reported cure for hiccups—working equally in unconscious people (presumably oblivious to punishment) with anesthetic-induced hiccup. Like many hiccup cures, this stimulates the vagus higher in its trajectory, interrupting the reflex.

One defective case does not invalidate a phenomenon. Surely other "psychogenic" hiccup reports rest on a sturdier foundation? A woman's hiccups were "psychogenic" because, it was announced, they were precipitated by an emotionally significant event. The touted trigger: her daughter's age—the age she had been when she herself was abused. (Hiccup is the obvious outcome.) Causal affirmation rested on a history of medical maladies triggered by emotionally significant events. A fall on ice was chalked up to an emotional event in the general temporal vicinity. Then there was her history of morbid fear of uterine cancer, so powerful it "caused" uterine bleeding, then led to uterine cancer itself. (The possibility that her fear of uterine cancer was justified—indeed, triggered by the abnormal uterine bleeding, which was actually due to the cancer that was later diagnosed—was not considered.)

In other instances, the psychogenic defense rested on cessation of hiccups with sleep. Proof positive. Except for pesky counterexamples a reading of the literature would expose. Like the boy whose recurrent hiccups initially resolved with sleep—but then didn't. And then his medullary brain tumor was diagnosed. (The medulla exerts tonic inhibition to the hiccup reflex; damage disrupts this inhibition.) A hiccup epidemic in a hospital ward was clearly mass psychogenic illness. Many contracted hiccups, so susceptibility to psychic contagion must have high penetrance. How then have friends, family, and hospital roommates of the many other persistent hiccup cases been spared? Might there be another explanation? How about actual contagion? *Streptococcus singultus* had caused epidemic hiccup in the past and could be passaged in rabbits, causing them to hiccup. No effort to hunt for such a cause was made. (*Singultus* is "hiccup" in medicalese.)

A review of this literature in days of yore (graduate school) revealed no report of psychogenic hiccup in which positive evidence corroborated a psychogenic cause. Worse, the foundation for psychogenic illness itself was supposition. There was no delineation of mechanisms by which such effects putatively occur, much less demand to prove such mechanisms were operating. Nor was there a clear exposition of what precisely was meant by "psychogenic," which morphs for the convenience of the expositor.

NOTE: I don't presume that physical ailments cannot have psychological triggers. Some "alternative medicine" approaches proffer putative means to discriminate which cases do, furnish testable hypotheses, and effect cures—a standard beyond that which "mainstream" medicine adopts.

Many psychogenic epithets have surrendered to evidence. Ulcers were psychogenic—until *Helicobacter pylori* and NSAIDS usurped the blame. So was most low back pain. By 1987, Finnish researcher Matti Joukamaa had it partly right: "Little is known about its [low back pain's] aetiology, its natural history and its treatment. This may explain why the myth exists that low back pain is often psychogenic."[ag] This prescience was undermined (or peer reviewers courted) when he added that those afflicted with back pain were, however, apt to harbor neuroses, and on top of that, weak egos—a revision, he proclaimed, of the prior view, in which conversion hysteria and psychosis dominated causes of back pain. (It's remarkable how the advent of workplace ergonomics helped gird weak egos.)

The newly minted "somatic symptom disorder" is the latest take on psychogenic illness, anointed in the last incarnation of the *Diagnostic and Statistical Manual.* (This is the tome that guides haruspication—I mean, psychiatry.) This dispenses with even the one requirement, lack of another cause, recognizing the pesky propensity for that lack to be sometimes—*horresco referens*—remedied, thus discrediting the doctor who declared the problem psychogenic. Now the doctor can skip the tiring pretense of actually looking for a cause, and if one is found anyway, he still saves face by virtue of the patient remaining impugned. The condition is "cured" only when the patient shuts up about her symptom(s). This helps the doctor and the health-care system. Never mind the patient.

The emperor never had clothes. The psychogenic designation is logically vacuous, not meaningfully defined, so not falsifiable, grounded in *petitio principii* (circular reasoning)—and functions as an assault. It impedes a search, when warranted, for legitimate conditions, breaches patient-doctor trust, effectively abandons the patient, and blames him for his affliction while also casting a pall of mental infirmity. It adds to (rather than mitigates) the patient's travails, antithetical to the dictum *primum non nocere*—first do no harm—that ought to guide medical care.

The psychogenic designation has long presumed that for any other condition a standard of evidence must be met. Yet for psychogenics, no standard is demanded: *Ipse dixit.* Proof by suggestion. Who could believe that? Someone who suffers from the delusion of psychogenic illness: It's all in the doctor's head.

CRIME ENTAILS ONLY THE ACTIONS OF CRIMINALS

EDUARDO SALCEDO-ALBARÁN
Philosopher; director, Scientific Vortex, Inc.

It sounds logical to say that to understand crime you must focus on criminals and felonies. But advances in social science give us reason to reconsider this idea.

Heroin is trafficked from Turkey across the Kapitan Andreevo security checkpoint in Bulgaria, to be sold in the richest countries of the European Union. More illegal drugs arrive in Europe from South America, via the countries of East Africa. In South Africa, racketeers, private security firms, and arms dealers conduct businesses together, erasing the boundary between legal and illegal financial transactions.

In Mexico, ferrous material, hydrocarbon condensate, and illegal drugs are trafficked and sold to legal and illegal firms and to individuals in the United States. *Los Zetas* and other criminal networks operating in Central America engage in human trafficking, kidnapping, and murdering of migrants before they cross the U.S. border. Between 2006 and 2010, some of those criminal networks laundered $881 million through a single legal bank inside the U.S. In 2012, the Criminal Division of the Department of Justice reported that the same bank "failed to monitor" $9.4 billion in the same period. Whoever has sent or received wire transfers, both inside and outside the U.S., will find it difficult to understand how one of the most important banks worldwide could "fail to monitor" $9.4 billion.

In all these cases, the participation of legitimate public servants, private citizens, and corporations is essential. In all these cases, bankers, attorneys, police, border officials, flight controllers, mayors, governors, presidents, and politicians co-opt and are co-opted by crime. Sometimes they're the instruments; sometimes they're the structural bridges connecting legality and illegality. They provide information, money, protection, knowledge, and social capital to criminal networks—a reason to define them as "unlawful" actors. However, they operate within legal agencies—a reason to define them as "lawful" actors. They seem lawful and unlawful at the same time. They're what we call "gray" actors, located and operating on the border between legality and illegality. They don't appear in the charts of criminal organizations, although they provide input for successful criminal operations.

Despite the significant role of these gray actors, social scientists interested in analyzing crime usually focus only on criminals and criminal actions. They tend to study crime through qualitative and quantitative data relating only to those "dark" elements, ignoring the fact that transnational and domestic crime is carried out by people who don't interact solely through criminal actions. This is a hypersimplified approach, because the dark elements are only the tip of the iceberg of global crime.

Their approach also assumes that society is a digital, binary system, in which the "good" and the "bad" guys, the "us" and "them," are perfectly distinguishable—a useful distinction in penal terms when simple algorithms ("If individual *X* executes action *Y*, then *X* is a criminal") orient judicial sentencing decisions. However, in sociological, anthropological, and psychological terms, this line is more difficult to define. If society is a digital system, it's certainly not a binary one.

This doesn't mean that crime is relative, or that we are all criminals because we're indirectly related to someone who commits a crime. It means only that defining and analyzing crime shouldn't consist of simple binary criteria like belonging to a group or committing a single action. Such criteria are useful applied to the boss of a criminal group or the specific act of shooting someone. But most of the time, affiliations and actions are complex and fuzzy; thus, as a society, we rely on and trust the intuition of a judge—someone who, despite simple algorithms, considers such factors as intention, context, and effect when pronouncing sentence. And this is why we aren't designing software for convicting criminals and passing sentences—such things are complicated.

Current tools for organizing and assessing large amounts of data are useful for understanding the complexity of crime. Formulating explicative models through social-network analysis or predictive models through machine learning that integrates several variables are examples of useful procedures. However, those procedures usually miss classic distinctions between "right" and "wrong," or ignore the fragmentation of scientific disciplines. Good and evil, right and wrong, legal and illegal—these are all context-driven concepts.

Economists, psychologists, anthropologists, and sociologists are often uncomfortable with the mixture of concepts required for analyzing complex behavior. Facing this complexity means integrating categories from many scientific domains, moving easily between macro and micro characteristics, even adopting new models of causality. This sounds like an impossible enterprise in traditional scientific realms. Social scientists are morally obliged to use the most accurate tools of observation when analyzing data and phenomena, because their observations inform the design and enforcement of policies. If inaccurate tools are used, bad decisions are made—as when a doctor diagnoses a tumor just by measuring body temperature. When the science we're studying is about understanding human trafficking, mass murders, or terrorism, using the best tools and providing the best inputs mean preserving lives.

It's therefore time to retire the idea that understanding crime means understanding the minds and actions of criminals. We must also retire other naïve concepts, such as "organized crime" and the idea that any current nation or government evolves without any criminal influence. These are nicely simplified concepts that work well in theoretical models in the classrooms and journals that manage to evade the complexity and vagueness of society. But if we don't deal with society's true complexity using the diverse tools provided by science, we'll have to deal with that complexity in the streets and the courtrooms—like it or not.

STATISTICAL SIGNIFICANCE

CHARLES SEIFE

Professor of journalism, New York University; former journalist, Science*; author,* Virtual Unreality

It's a boon for the mediocre, the credulous, the dishonest, and the merely incompetent. It turns a meaningless result into something publishable, transforms a waste of time and effort into the raw fuel of scientific careers. It was designed to help researchers distinguish a real effect from a statistical fluke, but it has become a quantitative justification for dressing nonsense up in the mantle of respectability. And it's the single biggest reason that most of the scientific and medical literature isn't worth the paper it's written on.

When used correctly, the concept of statistical significance is a measure to rule out the vagaries of chance—nothing more, nothing less. Say, for example, you're testing the effectiveness of a drug. Even if the compound is completely inert, there's a good chance (roughly 50 percent, in fact) that patients will respond better to your drug than to a placebo. Randomness alone might imbue your drug with seeming efficacy. But the more marked the difference between the drug and the placebo, the less likely it is that randomness alone is responsible. A "statistically significant" result is one that has passed an arbitrary threshold. In most social-science journals and the medical literature, an observation is typically considered statistically significant if there's less than a 5 percent chance that pure randomness can account for the effect you're seeing. In physics, the threshold is usually lower, often 0.3 percent (three sigma) or even

0.00003 percent (five sigma). But the essential dictum is the same: If your result is striking enough to pass that threshold, it's given a weighty label: "statistically significant."

Most of the time, though, the term isn't used correctly. If you look at a typical paper published in the peer-reviewed literature, you'll see that never is just a single observation tested for statistical significance but instead handfuls, or dozens, or even 100 or more. A researcher looking at a painkiller for arthritis sufferers will look at data to answer question after question: Does the drug help a patient's pain? Does it help a patient with knee pain? With back pain? With elbow pain? With severe pain? With moderate pain? With moderate to severe pain? Does it help a patient's range of motion? Quality of life? Each one of these questions is tested for statistical significance, and typically gauged against the industry-standard 5-percent rule. That is, there's a 5-percent chance—1 in 20—that randomness will make a worthless drug seem effective. But test ten questions, and there's a 40-percent chance that randomness will, indeed, deceive you when answering one or more of these questions. And the typical paper asks more than ten questions, often many more. It's possible to correct for this "multiple comparisons" problem mathematically (though it's not the norm to do so). It's also possible to fight this effect by committing to answer just one main question (though in practice such "primary outcomes" are surprisingly malleable). But even these corrections often can't take into account numerous effects that can undermine a researcher's calculations—such as how subtle changes in data classification can affect outcomes. (Is "severe" pain a 7 or above on a 10-point scale, or is it an 8 or above?) Sometimes these issues are overlooked; sometimes they're deliberately ignored or even manipulated.

In the best-case scenario, when statistical significance is calculated correctly, it doesn't tell you much. Sure, chance alone is (relatively) unlikely to be responsible for your observation. But it doesn't reveal anything about whether the protocol was set up correctly, whether a machine's calibration was off, whether a computer code was buggy, whether the experimenter properly blinded the data to prevent bias, whether the scientists truly understood all the possible sources of false signals, whether the glassware was properly sterilized, and so on. When an experiment fails, it's more than likely that the blame rests not on randomness—on statistical flukes—but instead on a good old-fashioned screwup somewhere.

When physicists at CERN claimed to have spotted neutrinos moving faster than light, a six-sigma level of statistical significance (and an exhaustive check for errors) wasn't enough to convince smart physicists that the CERN team had messed up somehow. The result clashed not only with physical law but with observations of neutrinos coming from supernova explosions. Sure enough, a few months later, the flaw (a subtle one) finally emerged, negating the team's conclusion.

Screwups are surprisingly common in science. Consider, for example, the fact that the Food and Drug Administration inspects a few hundred clinical laboratories each year. Roughly 5 percent of inspections come back with findings that the laboratory is engaged in "significant objectionable conditions and practices" so egregious that its data are considered unreliable. Often these practices include outright fraud. Those are just the blindingly obvious problems visible to an inspector; it's hard to imagine that the real number of lab screwups isn't double or triple or quintuple that. What value is there in calling something statistically significant at the 5-percent or 0.3-percent or even 0.00003-percent level if there's a 10-percent or 25-percent (or more) chance that the data is gravely undermined by a laboratory error? Even the most ironclad findings of statistical validity lose their meaning when dwarfed by the specter of error. Or worse yet, fraud.

Nevertheless, even though statisticians warn against the practice, a one-size-fits-all finding of statistical significance is all too often taken as a shortcut to determine whether or not an observation is credible or a finding is "publishable." As a consequence, the peer-reviewed literature is littered with statistically significant findings that are irreproducible and implausible—absurd observations with effect sizes orders of magnitude beyond what's even marginally believable.

The concept of "statistical significance" has become a quantitative crutch for the essentially qualitative process of whether or not to take a study seriously. Science would be much better off without it.

SCIENTIFIC INFERENCE VIA STATISTICAL RITUALS

GERD GIGERENZER

Psychologist; director, Center for Adaptive Behavior and Cognition, Max Planck Institute for Human Development, Berlin; author, Risk Savvy

As a young man, Gottfried Wilhelm Leibniz had a beautiful dream: to discover the calculus that could map every single idea in the world into symbols. Such a universal calculus would put an end to all scholarly bickering. Every passionate *Edge* discussion, for one, could be swiftly resolved by dispassionate calculation. Leibniz optimistically estimated that a few skilled people should be able to work the whole thing out in five years. Yet nobody, Leibniz included, has yet found that Holy Grail.

Nonetheless, Leibniz's dream is alive and thriving in the social and neurosciences. Because the object of the dream has not been found, ersatz objects serve in its place. In some fields it's multiple regression, in others Bayesian statistics. But the champ is the null ritual:

1. Set up a null hypothesis of "no mean difference" or "zero correlation." Don't specify the predictions of your own research hypothesis.
2. Use 5 percent as a convention for rejecting the null. If significant, accept your research hypothesis. Report the result as $p < .05$, $p < .01$, or $p. < .001$, whichever comes next to the obtained p-value.
3. Always perform this procedure.

Not for a minute should anyone think this procedure has much to do with statistics proper. Sir Ronald Fisher, to whom it has been wrongly attributed, in fact wrote that no researcher should use the same level of significance from experiment to experiment. The eminent statisticians Jerzy Neyman and Egon Pearson would roll over in their graves if they knew about its current use. Bayesians, too, have always detested p-values. Yet open any journal in psychology, business, or neuroscience and you're likely to encounter page after page with p-values. To give just a few illustrations: In 2012, the average number of p-values in the *Academy of Management Journal*, the flagship empirical journal in its field, was 116 per article, ranging between 19 and 536! Typical of management, you might think. But 89 percent of all behavioral, neuropsychological, and medical studies with humans published in 2011 in *Nature* reported p-values only—without considering effect size, confidence interval, power, or model estimation.

A ritual is a collective or solemn ceremony consisting of actions performed in a prescribed order. It typically involves sacred numbers or colors, delusions to avoid thinking about why one is performing the actions, and fear of being punished if one stops doing so. The null ritual contains all these features.

The number "5 percent" is held sacred, allegedly telling us the difference between a real effect and random noise. In fMRI studies, the numbers are replaced by colors, and the brain is said to light up.

The delusions are striking. If psychiatrists had any appreciation of statistics, they would have entered these aberrations into the *Diagnostic and Statistical Manual of Mental Disorders*. Studies in the U.S., U.K., and Germany show that most researchers don't (or don't want to) understand what a p-value means. They confuse it with the probability of a hypothesis—that is, p(Data|Ho) with p(Ho|Data)—or with some other bit of wishful thinking, such as the probability that the data can be replicated. Startling errors are published in top journals. For instance, an elementary point is that in order to investigate whether two means differ, we should test their difference. What shouldn't be done is to test each mean against a common baseline, such as: "Neural activity increased with training ($p < .05$) but not in the control group ($p > .05$)." A 2011 paper in *Nature Neuroscience* presented an analysis of neuroscience articles in *Science, Nature, Nature Neuroscience, Neuron,* and *The Journal of Neuroscience* and showed that although seventy-eight articles did as they should, seventy-nine used the incorrect procedure.

Not performing the ritual can provoke great anxiety, even when it makes absolutely no sense. In one study (the authors' names are irrelevant), Internet participants were asked whether there was a difference between heroism and altruism. The great majority felt so: 2,347 respondents (97.5 percent) said yes; 58 said no. What did the authors do with that information? They computed a chi-square test, calculated that c2(1) = 2178.60, $p < .0001$, and came to the astounding conclusion that there were indeed more people saying yes than no.

One manifestation of obsessive-compulsive disorder is the ritual of compulsive hand washing even if there is no reason to do so. Likewise, researchers adhering to the null ritual perform statistical inferences all the time, even in situations where there's no point—that is, when no random sample was taken from a population, or no population was defined in the first place. In those cases, the statistical model of repeated random sampling from a population doesn't even apply and good descriptive statistics is called for. So even if a significant p-value has been calculated, it's not clear what population is meant. The problem isn't statistics but its mistaken use as an automatic inference machine.

Finally, just as compulsive worrying and hand washing can interfere with the quality of life, the craving for significant p-values can undermine the quality of research. Which it has: Finding significant theories has been largely replaced by finding significant p-values. This surrogate goal encourages such questionable research practices as selectively reporting studies and conditions that "worked," or excluding data after looking at their effect on the results. According to a 2012 survey in *Psychological Science* of some 2,000 psychologists, over 90 percent admitted to having engaged in at least one of these or other questionable research practices. This massive borderline cheating in order to produce significant p-values is likely more harmful to progress than the rare cases of outright fraud. One harmful outcome is a flood of published but irreproducible results. Genetic and medical research using Big Data has encountered similar surprises when trying in vain to replicate published findings.

I don't mean to throw out the baby with the bathwater; statistics offers a highly useful toolbox for researchers. But it's time to get rid of statistical rituals that nurture automatic and mindless inferences. Scientists should study rituals, not perform rituals themselves.

THE POWER OF STATISTICS

EMANUEL DERMAN

Professor of financial engineering, Columbia University; former head, Quantitative Strategies Group, Equities Division, Goldman Sachs; author, Models.Behaving.Badly

I grew up among physicists, whose *modus operandi* is to observe the world, experiment with it, develop hypotheses and theories and models, suggest further experiments, and use statistics to analyze the results, thereby comparing mental imaginings with actual events. Statistics is simply their tool for confirmation or denial.

But nowadays the world, and especially the world of the social sciences, is increasingly in love with statistics and data science as a source of knowledge and truth itself. Some people have even claimed that computer-aided statistical analysis of patterns will replace our traditional methods of discovering the truth—not only in the social sciences and medicine but in the natural sciences, too.

We must be careful not to get too enamored of statistics and data science and thereby abandon the classical methods of discovering the great truths about nature (and man is nature, too). A good example of the classical power is Kepler's 17th-century discovery of his second law of planetary motion, which is in fact less a law than the recognition and description of a pattern. Kepler's second law states that the line between the sun and a moving planet sweeps out equal areas in equal times. This deep symmetry of planetary motion implies that the closer a planet is to the sun, the more rapidly it moves along its orbit.

But notice that *there is no line* between a planet and the sun. Kepler's still astonishing insight required examining Tycho Brahe's data, a long mental struggle, a burst of intuition—use an invisible line!—and then checking his hypothesis. Data, intuition, hypothesis, and finally comparison with data is the time-honored process.

Kepler's second law is in fact a statement of the conservation of angular momentum that followed later from Newton's theories of motion and gravitation. Newton's theories were readily and immediately accepted because Kepler's three verified laws could be derived from them. John Maynard Keynes wrote of Newton, 300 years later, "I fancy his preeminence is due to his muscles of intuition being the strongest and most enduring with which a man has ever been gifted."

Statistics—the field itself—is a kind of Caliban, sired somewhere on an island in the region between mathematics and the natural sciences. It's neither purely a language nor purely a science of the natural world but, rather, a collection of techniques to be applied to test hypotheses. Statistics in isolation can seek only to find past tendencies and correlations and assume they will persist. But in a famous unattributed phrase, correlation is not causation.

Science is a battle to find causes and explanations amid the confusion of data. Let us not get too enamored of data science, whose great triumphs so far are mainly in advertising and persuasion. Data alone have no voice. There are no "raw" data, as Kepler's saga shows. Choosing what data to collect and how to think about them requires insight into the invisible. Making good sense of the data collected requires the classic conservative methods: intuition, modeling, theorizing, and then, finally, statistics.

REPRODUCIBILITY

VICTORIA STODDEN
Computational legal scholar; assistant professor of statistics, Columbia University

I'm not talking about retiring the abstract idea or its place in scientific discourse and discovery; instead I'm suggesting redefining specifically what is meant by that word and using more appropriate terminology for the various research environments scientists work in.

When Robert Boyle brought the concept of reproducibility into scientific discourse in the 1660s, what comprised scientific experimentation and discovery was twofold: deductive reasoning, such as mathematics and logic, and Francis Bacon's relatively new machinery of induction. How to verify correctness was well established in logical deductive systems, but verifying experimentation was much harder. Through his attempts with Robert Hooke to establish a vacuum chamber, Boyle made a case that inductive, or empirical, findings—those arising from observing nature and then drawing conclusions—must be verified by independent replication. It was then that empirical research came to be published with sufficient detail regarding procedure, protocols, equipment, and observations such that other researchers could repeat the procedure and presumably therefore the results.

This conversation is complicated by today's pervasive use of computational methods. Computers are unlike any previous scientific apparatus, because they act as a platform for the implementation of a method rather than directly as an instrument. This creates additional instructions to be communicated as part of Boyle's vision of replicable research—the code and digital data.

The communication gap hasn't gone unnoticed in the computational science community, and, somewhat reminiscent of Boyle's day, many voices are calling for new standards of scientific communication—this time to include digital scholarly objects such as data and code. Irreproducible computational results from genomics research at Duke University in recent years focused attention on this issue and led to a report by the Institute of Medicine of the National Academies recommending new standards for clinical trials' approval for computational tests arising from computational research.

The report recommended, for the first time, that software associated with a computational test be fixed at the beginning of the approval process and thereafter made "sustainably available." A subsequent workshop at Brown University on "Reproducibility in Computational and Experimental Mathematics" (of which I was a co-organizer) produced recommendations about the appropriate information to include when publishing computational findings, including access to code, data, and implementation details. Reproducibility in this context should be relabeled "computational reproducibility."

Computational reproducibility can then be distinguished from empirical reproducibility, or Boyle's version of the appropriate communication for noncomputational empirical scientific experiments. Making this distinction is important, because traditional empirical research is running into a credibility crisis of its own regarding replication. As Nobel laureate (and *Edgie*) Daniel Kahneman has noted in reference to the irreproducibility of certain psychological experiments, "I see a train wreck looming."

What's becoming clear is that science can no longer be relied on to generate "verifiable facts." In these cases, the discussion concerns empirical reproducibility rather than computational reproducibility. But calling both types "reproducibility" muddies the waters and confuses discussion aimed at establishing reproducibility as a standard. I believe there is (at least) one more distinct source of irreproducibility—"statistical reproducibility." Addressing issues of reproducibility through improvements to the research-dissemination process is important but insufficient.

We also need to consider new measures to assess the reliability and stability of statistical inferences, including developing new validation measures and expanding the field of uncertainty quantification to develop measures of statistical confidence and a better understanding of sources of error, especially when large multisource data sets or massive simulations are involved. We can also do a better job of detecting biases arising from statistical reporting conventions established in a data-scarce, pre-computational age.

A problem with any one of these three types of reproducibility—empirical, computational, and statistical—can be enough to derail the establishing of scientific facts. Each type calls for different remedies—improving existing communication standards and reporting (empirical reproducibility); making computational environments available for replication purposes (computational reproducibility); and the statistical assessment of repeated results for validation purposes (statistical reproducibility)—each with different implementations. These are broad suggestions, and each type of reproducibility can demand different actions depending on the details of the scientific research context, but confusing these very different aspects of the scientific method will slow our resolution of Boyle's old discussion that started with the vacuum chamber.

THE AVERAGE

NICHOLAS A. CHRISTAKIS

Goldman Family Professor of Social and Natural Science, Yale University; coauthor (with James H. Fowler), Connected

Ever since the landmark invention of diverse statistical techniques 100 years ago that let us properly compare the difference between the averages of two groups, we have deluded ourselves into thinking that it's such differences that are the salient—and often the only—important one between groups. We've spent a century observing and interpreting such differences. We've become almost obsessed, and we should stop.

Yes, we can reliably say that men are taller than women, on average; that Norwegians are richer than Swedes; that first-born children are smarter than second-born children. And we can do experiments to detect tiny differences in means—between groups exposed and unexposed to a virus, or between groups with and without a particular allele of a gene. But this is too simple and too narrow a view of the natural world.

Our focus on averages should be retired. Or, if not retired, given an extended vacation. During this vacation, we should catch up on another sort of difference between groups which has gotten short shrift: We should focus on comparing the difference in variance—which captures the spread, or range, of measured values—between groups.

Part of the reason we've focused so much on the average is that the statistical tools for computing and comparing averages are so much easier and well developed. It's much harder to compare whether the variance of one group is different from the variance of another. But this calls to mind the joke about the drunk searching for his keys on his knees under a lamp post because the light is better there. Drunk with statistical power, we've persuaded ourselves that the mean of a distribution is its most important property. But often it's not.

For example, we've focused on the differences in average wealth between groups—whether the United States is richer than other countries and what might have caused this, or whether bankers make more money than consultants and how this affects the professional choices of graduating college students. But the distribution of wealth in the groups may be equally important in explaining collective and individual outcomes and choices. Even if the U.S. and Sweden have the same average income (roughly speaking), the variance in income is much higher in the U.S. (income inequality is greater), and this fact, rather than any difference in means between the groups, may help explain what happens to people in these societies. For example, it may be better for the health of a group, and (on average!) for the health of the individuals within it, for the group to have a more equal distribution of income even if the average income is somewhat lower. We might wish for more equality at the expense of wealth.

Here's a hypothetical example leading to the opposite practical conclusion about inequality: When forming a crew of sailors for a sailboat, what would be best? To have all ten of the sailors have the same level of myopia, with mean vision of 20/200, or to have a group of sailors in which nine had even worse vision but one had perfect vision? The average vision could be the same in both groups, but for the purposes of sailing the boat effectively, and for the survival of all aboard, it might be better to have more rather than less inequality. We might wish for more inequality at the expense of vision.

Or consider a medical example of how variance is important: There may be two conditions with equal average prognoses—say, advanced AIDS and advanced liver cirrhosis—but doctors may offer "Do Not Resuscitate" orders to AIDS patients at much higher rates. It's tempting to conclude that doctors are more eager to avoid resuscitating AIDS patients, perhaps for discriminatory reasons. But the real reason may be that the variance in survival in the AIDS group is much higher and there may be many more patients in that group who will die imminently. The doctors may be oriented to this fact rather than to the average survival of the two groups; they may reason that they can wait to offer DNR orders to the cirrhosis patients.

A familiarity with variance would also allow us to make sense of the famously controversial hypothesis regarding why there are more male math professors at major universities: The mean overall math aptitude among men and women might be the same, but the variance in men might be higher. If so, this would mean that there are more men at the very bottom of the distribution (and, indeed, boys are roughly three times more likely to be mentally disabled than girls) but also that there are more men at the upper end of the distribution.

When we focus mainly on the mean, we miss the chance to observe interesting and important things about the world. And a restricted view has adverse practical as well as scientific implications. Do we want a richer, less equal society? Do we want educational programs to increase the equality of test scores, or the average? Will a cancer drug that makes some patients live longer and kills others sooner still be preferred by patients even if it has no effect on average survival? To really understand the relevant tradeoffs, we must acquire not only the tools but also the vision to focus on variance.

STANDARD DEVIATION

NASSIM NICHOLAS TALEB

Distinguished Professor of Risk Engineering, NYU Polytechnic School of Engineering; author, Incerto (Antifragile, The Black Swan, Fooled by Randomness, *and* the Bed of Procrustes)

The notion of standard deviation has confused hordes of scientists; it's time to retire it from common use and replace it with the more effective one of mean deviation. Standard deviation, STD, should be left to mathematicians, physicists, and mathematical statisticians deriving limit theorems. There's no scientific reason to use it in statistical investigations in the age of the computer, as it does more harm than good—particularly with the growing class of people in social science mechanistically applying statistical tools to scientific problems.

Say someone just asked you to measure the "average daily variations" for the temperature of your town (or the stock price of a company, or the blood pressure of your uncle) over the past five days. The five changes are: (-23, 7, -3, 20, -1). How do you do it?

Do you take every observation, square it, average the total, then take the square root? Or do you remove the sign and calculate the average? For there are serious differences between the two methods. The first produces an average of 15.7, the second 10.8. The first is technically called the root mean square deviation. The second is the mean absolute deviation, MAD. It corresponds to "real life" much better than the first—and to reality. In fact, whenever people make decisions after being supplied with the standard-deviation number, they act as if it were the expected mean deviation.

It's all due to a historical accident. in 1893, the great Karl Pearson introduced the term "standard deviation" for what had been known as "root mean square error." The confusion started then; people thought it meant "mean deviation." The idea stuck. Every time a newspaper has attempted to clarify the concept of market "volatility," it has defined it verbally as mean deviation yet produced the numerical measure of the (higher) standard deviation.

But it's not just journalists who fall for the mistake. I recall seeing official documents from the Department of Commerce and the Federal Reserve partaking of the conflation, even regulators in statements on market volatility. What's worse, Daniel Goldstein and I found that a high number of data scientists (many with PhDs) also get confused in real life.

It all comes from bad terminology for something nonintuitive. By a psychological bias Danny Kahneman calls "attribute substitution," some people mistake MAD for STD because the former is easier to come to mind.

(1) MAD is more accurate in sample measurements and less volatile than STD since it's a natural weight, whereas standard deviation uses the observation itself as its own weight, imparting large weights to large observations, thus overweighing tail events.

(2) We often use STD in equations but really end up reconverting it in the process into MAD (say, in finance, for option pricing). In the Gaussian world, STD is about 1.25 times MAD—that is, the square root of (Pi/2). But we adjust with stochastic volatility where STD is often as high as 1.6 times MAD.

(3) Many statistical phenomena and processes have "infinite variance" (say, the popular Pareto 80/20 rule) but have finite, and very well behaved, mean deviations. Whenever the mean exists, MAD exists. The reverse (infinite MAD and finite STD) is never true.

(4) Many economists have dismissed infinite-variance models, thinking these meant "infinite mean deviation." Sad, but true. When the great Benoit Mandelbrot proposed his infinite variance models fifty years ago, economists freaked out because of the conflation.

It is sad that such a minor point can lead to so much confusion. Our scientific tools are way too far ahead of our casual intuitions, which starts to be a problem with science. So I close with a statement by Sir Ronald A. Fisher: "The statistician cannot evade the responsibility for understanding the process he applies or recommends."

And the probability-related problems with social and biological science do not stop there: There are bigger problems with researchers using statistical notions out of a can without understanding them and babbling "*n* of 1" or "*n* large," or "This is anecdotal" (for a large Black Swan–style deviation), mistaking anecdotes for information and information for anecdote. The majority use regression in their papers in "prestigious" journals without quite knowing what it means and what claims can—and cannot—be made from it. Because of little check from reality and lack of skin-in-the-game, coupled with a fake layer of sophistication, social scientists can make elementary mistakes with probability yet continue to thrive professionally.

STATISTICAL INDEPENDENCE

BART KOSKO
Professor of electrical engineering and law, University of Southern California; author, Noise

It's time for science to retire the fiction of statistical independence.

The world is massively interconnected through causal chains. Gravity alone causally connects all objects with mass. The world is even more massively correlated with itself. It is a truism that statistical correlation doesn't imply causality. But it is a mathematical fact that statistical independence implies no correlation at all. None. Yet events routinely correlate with one another. The whole focus of most Big Data algorithms is to uncover just such correlations in ever larger data sets.

Statistical independence also underlies most modern statistical sampling techniques. It's often part of the very definition of a random sample. It underlies the old-school confidence intervals used in political polls and in some medical studies. It even underlies the distribution-free bootstraps or simulated data sets that increasingly replace those old-school techniques.

White noise is what statistical independence should sound like. The hisses and pops and crackles of true white-noise samples are all statistically independent of one another. This holds no matter how close the noise samples are in time. That means the frequency spectrum of white noise is flat across the entire spectrum. Such a process does not exist, because it would require infinite energy. That hasn't stopped generations of scientists and engineers from assuming that white noise contaminates measured signals and communications.

Real noise samples aren't independent. They correlate to some degree. Even the thermal noise bedeviling electronic circuits and radar devices has only an approximately flat frequency spectrum and then over only part of the spectrum. Real noise doesn't have a flat spectrum. Nor does it have infinite energy. So real noise is colored pink or brown or some other strained color metaphor that depends on how far the correlation reaches among the noise samples. Real noise is not and cannot be white.

A revealing problem is that there are few tests for statistical independence. Most tests tell at most whether two variables (not the data itself) are independent. And most scientists would be hard pressed to name even them. So the overwhelming common practice is simply to assume that sampled events are independent. Just assume that the data are white. Just assume that the data are not only from the same probability distribution but also statistically independent. An easy justification for this is that almost everyone else does it and it's in the textbooks. This assumption has to be one of the most widespread instances of groupthink in all of science.

The reason we so often assume statistical independence isn't its real-world accuracy. We assume statistical independence because of its armchair appeal: It makes the math easy. It often makes the intractable tractable. Statistical independence splits compound probabilities into products of individual probabilities. (Then often a logarithm converts the probability product into a sum, because it's easier still to work with sums than products). And it's far easier to lecture would-be gamblers that successive coin flips are independent than to conduct the fairly extensive experiments with conditional probabilities required to factually establish such a remarkable property. That holds because in general a compound or joint probability always splits into a product of conditional probabilities. The so-called multiplication rule guarantees this factorization. Independence further reduces the conditional probabilities to unconditional ones. Removing the conditioning removes the statistical dependency.

The Russian mathematician Andrei Markov made the first great advance over independence or whiteness when he studied events that statistically depend only on the immediate past. That was over a century ago. We still wrestle with the math of such Markov chains and find surprises. The Google search algorithm rests in large part on finding the equilibrium eigenvector of a finite Markov chain. The search model assumes that Internet surfers jump at random from Web page to Web page, much as a frog hops from lily pad to lily pad. The jumps and hops aren't statistically independent. But they are probabilistic. The next Web page you choose depends on the page you're now looking at. Real Web surfing may well involve probabilistic dependencies reaching back to several visited websites. It's a good bet that the human mind isn't a Markov process. Yet relaxing independence to even one-step or two-step Markov dependency has proved a powerful way to model diverse streams of data, from molecular diffusion to speech translation.

It takes work to go beyond the simple Markov property where the future depends only on the present and not on the past. But we have ever more powerful computers that do just such work. And many more insights will surely come from the brains of motivated theoreticians. Giving up the crutch of statistical independence can only spur more such results.

Science needs to take seriously its favorite answer: It depends.

CERTAINTY. ABSOLUTE TRUTH. EXACTITUDE.

RICHARD SAUL WURMAN
Founder, TED conference, e.g. conference, TEDMED conference; architect, cartographer; author, Information Architects

A wonderful diagram is the approximate theory of the sun-centered solar system of heliocentrism done by Nicolaus Copernicus, which he published in 1543.

It would never be published in any academic circles today, because it is not correct. The orbits aren't circular, they're elliptical; they're not all on the same plane, and the diagram is completely disproportionate and does not represent the distances between the planets or from the planets to the sun. It's a diagram of approximation. It's a diagram that gave permission for others, as Tycho Brahe released his documentation and his measurements so that Kepler could come up with a more approximate notion of our planetary universe incorporating more accurate geometries.

What I'm suggesting be retired are the three terms above, and what I suggest to be embraced is more academic leeway in theories of approximation that give permission for others to see and discover new patterns.

THE ILLUSION OF SCIENTIFIC PROGRESS

PAUL SAFFO

Technology forecaster; consulting associate professor, Stanford University

> Nature and nature's laws lay hid in night; God said "Let Newton be" and all was light. —Alexander Pope

The breathtaking advance of scientific discovery has the unknown on the run. Not so long ago, the Creation was 6,000 years old and Heaven hovered a few thousand miles above our heads. Now Earth is 4.5 billion years old and the observable universe spans 92 billion light-years. Pick any scientific field and the story is the same, with new discoveries—and new life-touching wonders—arriving almost daily. Like Pope, we marvel at how hidden Nature is revealed in scientific light.

Our growing corpus of scientific knowledge evokes Teilhard de Chardin's arresting metaphor of the noosphere, the growing sphere of human understanding and thought. In our optimism, this sphere is like an expanding bubble of light in the darkness of ignorance.

Our optimism leads us to focus on the contents of this sphere, but its surface is more important, for that's where knowledge ends and mystery begins. As our scientific knowledge expands, contact with the unknown grows as well. The result is not merely that we've mastered more knowledge (the sphere's volume) but that we've also encountered an ever expanding body of previously unimaginable mysteries. A century ago, astronomers wondered whether our galaxy constituted the entire universe; now they tell us we probably live in an archipelago of universes.

The science establishment justifies its existence with the big idea that it offers answers and ultimately solutions. But privately every scientist knows that what science really does is discover the profundity of our ignorance. The growing sphere of scientific knowledge is not Pope's night-dispelling light but a campfire glow in the gloom of vast mystery. Touting discoveries helps scientists to secure funding and gain tenure, but perhaps the time has come to retire discovery as the ultimate measure of scientific progress.

Let us measure progress not by what is discovered but rather by the growing list of mysteries that remind us of how little we really know.

NOTES

56 a. Alan H. Guth, "Eternal Inflation and its implications," arXiv:hep-th/0702178v1 22 Feb. 2007.

64 b. *Max Planck: Scientific Autobiography and Other Papers,* trans. Frank Gaynor (New York: Philosophical Library, 1949).

75 c. Letter to Richard Bentley, Feb. 25, 1693, quoted in Richard S. Westfall, *Never at Rest: A Biography of Isaac Newton* (Cambridge University Press, 1983), p. 505.

122 d. Community Planning Study: Snowmass 2013, "Energy Frontier Lepton and Photon Colliders."

123 e. *Engineering & Science*, Winter 1995.

153 f. Anthony R. Cashmore, "The Lucretian Swerve," *Proc. Nat. Acad. Sci.* 107:10, 4499-504; DOI10.1073/pnas.0915161107 (2010).

177 g. William F. N. Chan *et al.*, "Male Microchimerism in the Human Female Brain," *PLOS ONE*, Sept. 26, 2012. DOI: 10.1371/journal.pone.0045592.

178 h. Samir Zaidi et al., "*De novo* mutations in histone-modifying genes in congenital heart disease," *Nature* 498:7453 (2013).

178 i. James R. Lupski, "Genome Mosaicism—One Human, Multiple Genomes," *Science* 341, 358 (2013) DOI: 10.1126/science.1239503.

178 j. Maeve O'Huallachain et al., "Extensive genetic variation in somatic human tissues," *Proc. Nat. Acad. Sci.*, 109:44, 18018-23 (2012).

193 k. Ian C. G. Weaver et al., "Epigenetic programming by maternal behavior," *Nat. Neurosci.* 7:8, 847-54 (2004). doi: 10.1038/nn1276.

198 l. S. H. Rhee et al., "Early concern and disregard for others as predictors of antisocial behavior," *Jour. Child Psychol. & Psychiatry* 54, 157-66 (2013).

215 m. George E. Newman & Daylian M. Cain, "Tainted Altruism: When Doing Something Good Is Evaluated as Worse

Than Doing No Good at All," *Psychol. Sci.*, Jan. 8, 2014, doi: 10.1177 /0956797613504785.

217 n. Fritz Heider & Marianne Simmel, "An Experimental Study of Apparent Behavior," http://www.all-about-psychology.com/fritz-heider.html.

230 o. "Sleep and Obesity," *Curr. Opin. Clin. Nutr. Metab. Care* 14(4):402-12 (2011). doi: 10.1097/MCO.0b013e3283479109.

231 p. "Mice Fall Short as Test Subjects for Some of Humans' Deadly Ills," *New York Times*, February 11, 2013.

231 q. *The Desminopathy Reporter*, http://www.desminopathy.info/weblog/are-mice-useless-models-for.html.

232 r. Clinton Leaf, "Why we're losing the war on cancer," CNN.com, Jan. 12, 2007.

268 s. David Deutsch, "Philosophy will be the key that unlocks artificial intelligence," theguardian.com, 3 October 2012.

315 t. http://www.cjr.org/overload/interview_with_clay_shirky_par.php?page=all.

331 u. *An Essay on the Principle of Population*, Book IV, Chapter V.

407 v. " 'From Jerusalem to Jericho': A Study of Situational and Dispositional Variables in Helping Behavior," *Jour. Pers. & Soc. Psych.* 27:1, 100-108 (1973).

409 w. Solomon E. Asch, "Opinions and Social Pressure," *Sci. Amer.*, 193:5, p. 34 (1955).

410 x. Bert H. Hodges & Anne L. Geyer, "A Nonconformist Account of the Asch Experiments: Values, Pragmatics, and Moral Dilemmas," *Pers. & Soc. Psych. Rev.* 10:1, 4 (2006).

411 y. J. C. Bohorquez, et al., "Common Ecology Quantifies Human Insurgency," *Nature* 462:7275, 911-14 (2009). doi:10.1038/nature08631.

418 z. Eric Kandel, "The New Science of Mind," *New York Times*, Sept. 6, 2013.

438 aa. "The Weirdest People in the World?" *Behav. & Brain Sci.* (2010) doi: 10.1017/S0140525X0999152X.

444 ab. W. D. Hamilton, "The genetical evolution of social behaviour." I, *Jour. Theor. Biol.* 7(1):1-16 (1964).

449 ac. John Maynard Smith & Eörs Szathmáry, *The Major Transitions in Evolution* (New York: Oxford University Press), p. 7.

466 ad. http://www.theatlantic.com/magazine/archive/2010/01/the-catastrophist/307820/.

477 ae. Matthew W. Ohland et al., "Identifying and Removing a Calculus Prerequisite as a Bottleneck in Clemson's General Engineering Curriculum," *Jour. Engineering Educ.*, pp. 253-7, July 2004.

507 af. *Mental Pathology and Therapeutics*, 2nd ed.

513 ag. M. Joukamaa, "Psychological factors in low back pain," *Ann. Clin. Res.*, 19(2): 129-34 (1987).

INDEX

H

Q

R

BOOKS BY JOHN BROCKMAN